FOOD AND SOCIETY: A SOCIOLOGICAL APPROACH

THE REYNOLDS SERIES IN SOCIOLOGY

Larry T. Reynolds, *Editor*

by **GENERAL HALL, INC.**

FOOD AND SOCIETY: A SOCIOLOGICAL APPROACH

William C. Whit
Grand Valley State University

GENERAL HALL, INC.
Publishers
5 Talon Way
Dix Hills, New York 11746

FOOD AND SOCIETY:
A SOCIOLOGICAL APPROACH

GENERAL HALL, INC.
5 Talon Way
Dix Hills, New York 11746

Publisher: Ravi Mehra
Composition: Graphics Division, General Hall, Inc.

LIBRARY OF CONGRESS CATALOG CARD NUMBER: **95–78981**

ISBN: 1–882289–36–6 [paper]
1–882289–37–4 [cloth]

Manufactured in the United States of America

This book is dedicated to two people whose friendship and examples of social activism have sustained me over many years. My late cousin Charles Hill (1941–1990) practiced law in the interest of low-income people in Washington, D.C. And my close friend Phillip Villaire (1936–1993) was my model and teacher about nature, farming and living outside the academic world. He was a proofreader of this manuscript.

Contents

Preface

In 1986 the food anthropologist Mary Douglas badgered me for being uninformed about her newest book. I replied that there was virtually no tradition of research about food in sociology. I was starting from scratch.

Subsequently I have discovered that to be partially untrue. In 1939 W. Lloyd Warner directed a Culture and Foodways Project in southern Illinois. It emphasized socioeconomic variables and applied a mixture of Weberian and Durkheimian perspectives (Montgomery and Bennett 1979, 129). Additionally, in the early 1940s, the Committee on Food Habits (CFH) in the National Research Council utilized Kurt Lewin's 1943 study of the role of the 'gatekeeper" in influencing family food consumption patterns (Montgomery and Bennett 1979, 131).

The next significant 'sociological" work on food was that of Mary Douglas (1972) and Michael Nocod (1974) in England. Both dealt with the patterns and context of daily meals. Also in the early 1970s Christine Wilson published a '595-item annotated bibliography of food habits" (Montgomery and Bennett 1979, 134).

The year 1983 marked the appearance of Ann Murcott's *The Sociology of Food and Eating*. This small edited volume raised for the first time the possibility of the development of a sociology of food. At approximately the same time, Dr. Alex McIntosh found himself assigned the food section in a U.S. Agency for International Development (AID) study on indicators of social development. Since then, he has taught and studied food and society issues at Texas A & M University. Subsequently Jeff Sobal (now at Cornell University) began a series of studies on obesity, a subarea of the sociology of food and nutrition.

In 1985 the first text, *The Cultural Feast: An Introduction to Food and Society*, appeared. Co-authoring were two anthropologists, Carol Bryant and Kathleen DeWalt, and two dietitians, Anita Courtney and Barbara Markesbery. In 1986 a British health educator, Paul Fieldhouse, entered the discussion with his *Food and Nutrition: Customs and Culture*. It combined a plethora of societally related food studies.

In 1986 Yvonne Lockwood (from food and folklore at Michigan State University) and I edited *Food and Society: A Collection of Syllabi and Instructional Materials*. Carried by the Association for the Study of Food and Society and by the American Sociological Association, this book has been acquired by more than 200 academics and public-service organizations.

The year 1987 marked the first meeting of the Association for the Study of Food and Society. Since then, I have been accumulating the notes and studies that form the basis of this book. It is the first publication to take a self-consciously sociological approach. And it is among the first works on the subject written by a sociologist.

Within sociology, I take a self-conscious interdisciplinary approach. I have always felt that the best sociologist takes conceptual material from other social sciences and the humanities and applies a sociological perspective to it.

I hope this undertaking provides the reader with the kind of *erklarung* experience that, in some way, alters his or her perspective on the world in which we live. Looking at the world and societies from a food perspective is both enlightening and intellectually satisfying. It has formed my intellectual perspective for the study of sociology for most of the last decade.

References

Bryant, Carol A., Anita Courtney, Barbara A. Markesbery, and Kathleen DeWalt. 1985. *The Cultural Feast: An Introduction To Food and Society*. New York: West Publishing Company.

Fieldhouse, Paul. 1986. *Food and Nutrition: Customs and Culture*. New York: Croom Helm.

McIntosh, Alex. 1975. *Social Indicators for the Monitoring of the Nutritional Aspects of Social Well Being*. Ph.D. dissertation. Iowa State University, Ames, IA.

Montgomery, Edward and John W. Bennett. 1979. 'Anthropological Studies of Food and Nutrition: The 1940s and 1970s." *Uses of Anthropology 11*: 124–44.

Murcott, Anne, ed. 1984. *The Sociology of Food and Eating*. Hants, England: Gower.

Whit, Bill, and Yvonne Lockwood. 1990. *Teaching Food and Society: A Collection of Syllabi and Instructional Materials*. College Station, Tex.: Association for the Study of Food and Society.

Acknowledgements

Every book is a collective effort. I have utilized the skills of an excellent undergraduate and graduate education. Dominant among all influences of graduate school is Dr. Bernard Phillips. In the early 1970s at Boston University's Sociology Department, Dr. Phillips excelled in both his ability to trust independent research and his encouragement to broaden the horizons of sociology. That I have chosen to explore an area new to sociology is a tribute to his original encouragement.

I owe thanks to Dr. Frank Bruno who, as academic vice president at Aquinas College, provided underwriting for the first annual meeting of the Association for the Study of Food and Society. And later, President Peter O'Connor approved an arrangement that allowed me to purchase a semester off from teaching to work on this book. I also want to thank the administration of Aquinas College for its support.

Enormous thanks are due to Judi Creamer who, in the midst of work for an entire building of faculty, managed to type and revise the manuscript for class use and for final copy.

In the context of a general-purpose undergraduate library, I want to thank the staff who ordered numerous food books and expedited interlibrary loan requests to meet my schedule. Included are Pam Luebke, Sr. Rose Marie Martin, O.P., and Jeannie Tuszynski.

I am especially grateful to Dr. Larry Reynolds, my editor, without whose request to undertake this book, I probably never would have done it. His advice and editorial skills sustained me through the extended period of writing and he performed a final editing. I would also like to thank Irene Glynn and Avarama Gingold for copyediting this manuscript.

My sincere thanks to the three readers of the first draft: Dr. Alex McIntosh of Texas A & M University, Dr. Laura DeLind of Michigan State University, and Dr. Jeff Sobal of Cornell University. Their critical comments formed the basis for the third and final draft.

I want to express my gratitude to the three proofreaders who took time out of busy schedules. They are Dr. Mitchell Kalpakgian of the English Department at Simpson College (and my college roommate of 1960 and 1962), Mrs. Barbara Hill of Milwaukee (my cousin), and Dr. Roger Nemeth, a sociologist at Hope College.

Finally, for providing emotional sustenance, I deeply appreciate the support of my large circle of friends and fellow faculty at Aquinas College. In addition, I received sustained support from the Villaire family, Dr. Eunice Thurman, and Ms. Carol Hill. On a daily basis, I need to show appreciation to my feline companion, Cheng Shu, who snoozes next to my elbow even as I write this page.

In any endeavor this large there will inevitably be errors. I take responsibility for those as well as for the creativity and analytical contribution this work makes to the field of food and society.

Introduction

> In no other realm of life do human beings expend so much energy, nor do we engage in so many interrelated activities. And in no other areas of our experience, perhaps, are we compelled to express so many aspects of our total humanity. Virtually nothing else we do in our daily lives speaks so loudly of our sense of art, aesthetics, creativity, symbolism, community, social propriety, and celebration as do our food habits and eating behavior.
>
> —Amy Schulman, 1983

The importance of food is undeniable. Yet, because it is so close and so obvious, we often fail to pay attention to it. This introductory text attempts to make explicit the many facts, thoughts, feelings, and rules about food that pervade our *lebenswelt* ("lived world"). The goal is to develop a multilevel, multidisciplinary approach to the interrelationships of food and society.

The text is organized from the experiences of the consumption of food through its preparation, distribution, storage, and production. We start with consumption—those food experiences closest to the lives of readers. And we seek to trace the story of food progressively back toward its production.

The text does not present a linear story of food. Instead, after an initial chapter designed to catch the reader's interest, we digress to the role of food in past societies. And we digress again to a summary analysis of the basics of nutrition. Each digression provides needed background for what is to follow.

The remainder of the text examines contemporary issues including body size, food and culture, food and society, food production, world hunger, and food innovations/successes. By the end of the text the reader should have an appreciation and understanding of the multifaceted world of food studies.

Chapter 1 raises certain existential issues that should excite reader interest in the social implications of the food we eat. It covers four basic subject areas: chicken, food cooperatives, rice, and breakfast cereals.

The production of chicken, our most common 'white" meat, is pregnant with examples that implicate the entire food system. Most chicken production is part of an economic system that is highly concentrated in a limited number of major producers. Workers are often severely exploited. And government regulation on the safety of chicken from these producers is often inadequate to guarantee the health of consumers. Chicken producers would prefer to irradiate chicken rather than attempt to raise a more healthy (and more expensive) bird.

In contrast, a series of food cooperatives across the country advocate low technology and economic diversity. Much of the food they sell is secured from local producers and organic farmers. Fewer industrial chemicals are used in food production, and there is a sense of fellowship in the smallness and democratic organization of neighborhood food co-ops.

Perhaps the most symbolic food in the past has been rice. As an example of the manner in which a crop can structure a civilization, rice-raising societies often are associated with large-scale bureaucratic organizations. Producing rice is an enterprise that requires cooperation in the regulation of water, seed, and fertilizer. Asian societies often evidence the high degree of organization necessary to grow rice.

At the other extreme are breakfast cereals. While they began as health foods, the economic concentration under which they are now produced leads to a form of economic competition that depends on advertising and product proliferation, not the original capitalist aspects of quality and price. Marketing and profit take precedence over people's need for healthy food.

All these subjects raise important issues about our contemporary food situation. But, it is important to recall the story of food from the beginning of human civilization so that we can gain an appreciation of past production and consumption patterns.

In chapter 2 we learn that history has provided many answers to the questions of production and consumption of food. We choose as our periodization: hunting and gathering, food-growing, and capitalist industrial societies. We tend to focus on the important and often neglected role of women in food production, preparation, and consumption.

In hunting and gathering societies, women generally provided the bulk of consumable food. Their gathering activity and participation in small hunts and trapping provided the basics on which could be added the episodic hunting of larger animals. This consumption pattern demonstrates the dominance of a basically vegetarian diet. But there is, of course, a great variation between different hunting and gathering societies.

Food-growing societies divide into gardening (hoe technology) and farming (plow technology). Both evidence a class structure that divides a ruling elite from the mass of people. Food consumption follows this pattern with the elite eating much meat and the mass (90 percent or more) eating a predominantly vegetarian diet.

How can one eat a vegetarian diet and get adequate nutrition? Chapters 3 and 4 provide the necessary scientific knowledge about nutrition. From the perspective of food history, the reader can judge the 'fit" between contemporary food theories and the groups that survived to tell their food stories.

Chapter 3 provides the basics of protein, fat, carbohydrates, vitamins and minerals. Most successful societies evolved a cuisine that incorporated adequate amounts of each. But contemporary societies must also include the role of petrochemicals as they have penetrated the food supply. Governments are often too beholden to concentrated business interests to regulate residues of pesticides, herbicides, growth hormones, and antibiotics adequately.

Chapter 4 uses a topical approach to illustrate the conflicting interests of the population (health) versus investors and business (profit) in food. The evolution of sugar use provides a paradigmatic example of the role of food in the generation of profit versus health. Sugar was part of the second triangular trade. Its world use was built on the slave trade, and both helped finance the capitalist industrial revolution. Sugar was the first junk food, and it continues in most of its contemporary uses in this tradition.

Meat is one of class divisions in various societies. After the age of hunting and gathering societies, meat was the prerogative of the elite. The mass of people ate very little meat, depending on diets primarily of vegetables, fruits, and grains. In our capitalist industrial society, meat has spread to the working and lower classes. In fast-food hamburgers, everyone partakes of the pervasive chemicalized meat of advanced industrial society. But our meat consumption comes at the cost of world ecological destruction. Rain forests are destroyed to make more grazing room for cattle. Peasants are pushed off their traditional land and deprived of their food source.

Ironically it is now the educated elite who are most likely, following the advice of scientific nutritionists, to cut back on meat. Today's elite often eats much like earlier hunters and gatherers. They 'graze." They eat predominantly vegetarian diets. And they may eat superior range-fed or organic meats.

But most of modern society eats the food that is so heavily advertised by the corporate interests that dominate commercial food production. In chapter 5 we see the results of modern diets: obesity and eating disorders. There are perhaps no more pressing concerns to college women today than issues of weight—obesity and thinness. Previous to 1920 in the United States, the rich set the standards of beauty with their wide girths. To be thin was to be avoided, a tradition that stretched back to the inadequate nutrition of the poor. Fatness was highly fashionable and desirable. With the flappers of the 1920s, however, came the first equation of thinness with beauty. While the working class fattened on sugar and fat-laden meat, the rich were beginning to admire the thin.

In contemporary America, women are caught in a dilemma: they want to eat the fattening (and profitable) foods so heavily advertised, yet stay as thin as the models who advertise them. The eating disorder bulimia is the incarnation of both social forces. Women eat all they want and then disgorge themselves to stay thin. A partial line can be drawn from the Roman food orgies with their adjacent vomitorium to contemporary eating disorders.

Foremost are questions of cultural values and norms and their relationship to the social order. Chapter 6 explores the culture of food, and chapter 7 its social order. Both raise the issue of idealist versus materialist explanations of culture and society. Are culture and values totally autonomous? Or are they evidence of an original *adaptive response* to material circumstances?

Chapter 6 uses Hindu cow proscriptions, Jewish food prohibitions, American Thanksgiving celebrations, Indian and Chinese cuisine, and American fast food as illustrations of the relative importance of idealist and materialist perspectives. Of these various customs, only Thanksgiving provides material that is more germane to idealist than materialist analysis. The rest are more clearly *adaptive responses* to material circumstances. Worst among them is the valuation of fast food. It provides a potentially nutricidal adaptation to the economic imperatives of capitalist industrially produced mass food. And the manner of its production raises ecological questions about the use of land, rain forests, and grain.

Chapter 7 explores the manner in which food is often best studied as an indicator of solidarity, group unity, socialization, and class/status. While food (in a manner similar to alcohol) may be a lubricant and an indicator of solidarity within a group, the refusal to eat with someone or some group is often an indicator of alien social relations between groups. Just as people often prepare their best meal for their best friends, they often refuse to eat with their enemies.

The occasion of eating can also be the method by which new members (often the young) are socialized to the values and norms of a society. In this context, the breakdown of the family meal in America often results from the institutionalization of industrially produced fast food and fast schedules.

Because people at the low end of the status (class) scale often emulate people further up, there now exists a situation where low-income people use malnutritive fast food as an emulation of the middle-class's fascination with beef. Even institutions that feed the poor and homeless often succumb to the norms of fried beef rather than more nutritional combinations of vegetarian dishes which could be prepared more cheaply. This raises many serious questions about the world's food production systems.

Chapter 8 explores food production and the world agricultural system. Once again, multinational corporate interests dominate societal needs. Giant corporate farms drive family farmers out of business at the rate of 2,000 per week. Corporate farming incorporates high-energy-intensive farming methods, including petrochemicals, large machinery, and hybrid seeds.

The Green Revolution used breeding to create new chemically dependent productive strains of plants. The biotechnological revolution utilizes genetic engineering with a system of patenting new life forms to assure corporate profit.

Unfortunately, both high-technology forms of farming come at large environmental costs. Most require so much water that water tables are dropping precipitously in many parts of the world. The remaining water often contains the runoff of petrochemicals. Soil is depleted and abandoned. Multitudes of crop varieties are eliminated, leading to the danger that humankind will no longer be able to draw from the biodiversity with which we started. And peasant subsistence farming is increasingly being destroyed in the search for more profitable crops.

Elements in contemporary socialist societies provide some alternatives. Most important is China with its low technology and intensive agriculture. Should China reach a level where it can produce its own biotechnology, one might expect developments that benefit people's general nutritional health rather than ones that are most financially profitable to international corporate interests.

In America, there are a few hopeful examples of farming in the interest of everyone. Wendell Berry (1988) cites the Amish manner of farming as the future of the family farm. Wes Jackson (1980) is doing experiments with no-till farming and harvesting perennials. And the Rodale Organization (1986) is conducting many experiments with scientific low and organic input farming.

But with all this agricultural technology at our disposal, why is there still mass hunger in the midst of a world that produces more than enough grain to make everyone fat? The answer lies in a world system where profit is more important than feeding people. Grain is fed to cows to produce beef for the affluent, rather than given directly to the hungry. People starve while the elite of their country export grain for private profit.

Chapter 9 attempts to make some sense of this situation by presenting a typological analysis of treatments of world hunger. Conservatives see world hunger as a problem of the values of the poor and give thanks for its Malthusian function as population control. Liberals see salvation in high-technological solutions and manipulation of governmental polices. Only socialist countries have actually 'solved" hunger by providing approximately equal access to food for most of their populations.

Chapter 10 explores some of the successes. Many are modern socialist societies. China, Cuba, North Korea, and the socialist state in India, Kerala, provide examples of situations where everyone has food. All assume a *right* to eat. In these countries, there has been some version of land reform, government planning, and "appropriate" technology.

In the developed world there remains a counterculture that seeks alternatives to the oligopolized food supply. Food co-ops, organic farming, farmers' markets, and contemporary nutritional knowledge form niches in a system that generally produces only the most profitable and mostly unhealthy food. Whether this "health" constituency can provide enough impetus for a change toward a nutritionally oriented food system is doubtful. But one would hope there will remain areas in the larger world economy where people and countries can place food first in their scheme of development. To do less would doom us to nutricide in the current world farming and eating system.

I hope that readers will find these chapters stimulating. Each could be a separate course. And I expect, in time, there will be graduate departments where one can study the interrelationship between food and society more intensively. Bon appetit!

References

Berry, Wendell. 1986. "In Defense of the Family Farm." Speech given at Kalamazoo College, Kalamazoo, Mich., October 8.

Jackson, Wes. 1980. *New Roots for Agriculture.* Lincoln: University of Nebraska Press.

Schulman, Amy. 1983. "The Rhetoric of Portions." In *Food and Eating Habits: Directions for Research*, ed. Michael Jones. Los Angeles: Folklore Society.

Chapter 1

Encountering Food

"Chicken Little"

European settlers brought to North America the tradition that eating fowl was special — behavior both ceremonial and festive. Rich families ate roast chicken, shared out by the knife-wielding chief male of the family on Sundays. King Henri IV of France had, in the sixteenth century, pronounced the ideal of extending this universally desired luxury to everybody: "I hope to make France so prosperous that every peasant will have a chicken in his pot on Sundays."

—Margaret Visser, 1986

Nutritionist Joan Gussow describes a scenario of chicken in the future: ''Chicken Little,' a legless, wingless, headless, featherless technological triumph, a giant mass of flesh 'fed by dozens of pipes' from which daily slices are cut to feed a populace otherwise reduced to soyaburgers" (Pohl and Kornbluth 1952, 1985, cited in Gussow, 1991).

From medieval France to the present, chicken has been a constant food. "In 1928 the Republican Party won an election in the United States with the help of the slogan 'A chicken in every pot'" (Visser 1986,116). Among today's health-conscious consumers, chicken (without its skin) has become popular as a 'light" meat. It is the food of choice for banquets and dinner parties because, for one reason or another, all the other choices have become too controversial.

Older Americans, including this author, can remember family farmers generally keeping some chickens for eggs and meat. In this low-technology world, eggs were collected manually and chickens were butchered by chopping off their heads and letting the remaining body struggle around until it lost life. The process was not particularly pretty. But no one had any illusions about it. The feathers were picked off by hand, and the chicken was cooked in any one of a variety of ways for consumption in the home of the family farm.

Furthermore, chickens fit into a farming niche that made ample use of their natural abilities. As Visser writes:

> The "corn cycle," which used to operate on many American farms, began with the feeding of whole kernels to cattle. Unchewed corn passed through the cows and into their excrement. Hogs scavenged in the dung, and ate the corn they found in it; sharp-eyed chickens were then allowed to hunt through the pig manure for particles of undigested corn. European visitors were amazed at American ingenuity at the saving of waste and labor. There was no grinding of corn and no need for feed containers. The hen took refuse and turned it into gold. (1986, 134)

But this is hardly the technology that produces Chicken McNuggets or other chicken products. Today's "scientific" approach to chicken raising partakes of all the manipulation of nature possible by corporate entities seeking to maximize profits. Perhaps the eventual outcome of this process will fulfill Gussow's description of Chicken Little.

In today's English "battery system," "laying chickens . . . live in small cages made of wire so that their droppings fall through the cage floor and can be removed from below with little difficulty" (Visser 1986, 136). Their raising is characterized by specially adjusted lighting and heating to produce maximum egg laying and/or meat development (Visser 1986, 134; Hall 1983, 13). Their food supply is no longer the waste of the barnyard but specially selected grains laced with both the crop chemicals (pesticides, herbicides) and added growth hormones and antibiotics to assure health and fast growth. The dangers to humans consuming these chemicals are yet to be fully determined. At this point we simply do not know what twenty years of chicken chemicals will do to us.

The technology of "processing" involves the application of industrial techniques to chicken killing and dissecting. In technologically sophisticated factories, usually employing more than 500 workers, chickens are first hung on shackles at the rate of twenty-five birds per minute. They are then "stunned" electrically from the metal shackle, their necks cut by machine, and the blood drained. "This process is designed to be quick and bloody, to keep [the] bird from tensing up and toughening its meat" (Hall 1989, 19). Labor exploitation is routine:

> Companies bossed by white men seem oblivious to a work force composed mostly of black women. They worry about stress on the chickens, yet devote hardly any attention to workers' disorders caused by the stress of keeping up with a production line moving at 90 birds a minute. . . .
>
> In truth, the 150,000 workers in poultry processing plants suffer one of the highest rates of injury and illness in American manufacturing . . . twice that of textile or tobacco workers and even higher than miners (Hall 1989, 15).

Pay is reasonable, but the workers are often physically incarcerated in the factories. White supervisors solicit sexual favors, and "work rules in some plants are so rigid that employees forbidden to take a break have urinated, vomited, and even miscarried while standing on the assembly line" (Goldofas 1989, 25).

Job injuries include "... eye infections, skin rashes, warts, and cuts" (Goldofas 1989, 25). Most common and debilitating is carpal tunnel syndrome. And even when workers can afford to get operations, the associated difficulties of working and loss of feeling in the hands often continue. "Workers who kill and process chickens—from white women in the Ozarks to black women in the Deep South—describe their work as 'modern slavery'" (Bates and Hall 1989, 11).

Government regulation is minimal. The U.S. Department of Agriculture's current policies mostly involve reprivatization, "turning federal responsibility for food safety over to the industry itself" (Devine 1989, 40). Food crusader Dr. Carl Telleen has documented that "carcasses contaminated with feces . . . are now simply rinsed with chlorinated water to remove the stain" (Devine 1989, 40). And "thousands of dirty chickens are bathed together in a chill tank, creating a mixture known as 'fecal soup' that spreads contamination from bird to bird" (Devine 1989, 40). The degree of contamination levels now hover routinely around 60 percent.

It is no wonder, then, that current government proposals advocate irradiation of chicken to kill bacteria. This poses the usual problems of the release of radiation due to human accidents. In addition, irradiation lessens some of the food value. The Center For Science in the Public Interest reports that irradiation changes the molecular structure and "reduces the vitamin C content of potatoes by up to 50%" (Jacobson 1992, 2). According to Jacobson, there is a cheaper and safer method. Dipping chickens in trisodium phosphate (TSP) is both safe and kills most of the disease-carrying bacteria (1992, 3).

Chicken-production ownership patterns are, like most of the food industry, becoming oligopolized: "More than a third of the chickens Americans eat come from four companies—Tyson, Agra, Gold Kist, and Holly Farms" (Hall 1989, 16). As with the rest of the food industry, vertical integration with contract farmers, feed mills, hatcheries, trucking companies, processors, and marketers has increased the size of the industry immensely. And in our political and economic system it is no surprise that the lessening of government supervision of poultry production comes as the result of corporate lobbying.

Finally, what has happened to the culture associated with chicken consumption? What happened to a "chicken in every pot"? Chickens, in their preindustrial state, have provided many cultural language metaphors. People have talked about the ego of the "cock" and about "pecking orders" (from the barnyard). Chickens have been birds of religious sacrifice. Cock fighting has been entertainment for centuries. In China and Thailand chickens have been used for divination. They have been sources of food on long sea voyages. And even today we can recall the use of chickens as an animate alarm clock. The accessibility of poultry to the common person has been one of the benefits of industrial culture.

What of the future? Are we, indeed, to be the consumers of biotechnologically produced blobs of chicken flesh fed by pipes and tubes and bearing no recognizable features of the chickens of the preindustrial, prescientific age? Alternative food cultures ask what kind of chicken we will eat. What is available to us besides the oligopolistic, impersonal, and sanitized chain store?

The Food Co-op

Imagine yourself entering a food cooperative. During an initial meeting in my Food and Society course, I take the class to the Eastown Food Co-op in Grand Rapids, Michigan. It has almost 5,000 members and has been in operation eighteen years. It occupies about 1,500 square feet of floor space in an "L" arrangement.

Many of the items are very different from a chain supermarket. As we peruse the crowded aisles, we find plexiglas bins of millet, six kinds of rice, dried split yellow peas, bulgar, and couscous. There are undyed cheeses. Basic grains include stone-ground, organic, hard and soft whole wheat flour (in clean, galvanized "garbage cans"). There is dried soy milk, soy flour, organic and nonorganic raisins and dates, noninstant dried milk (and instant as well), and a wide selection of organic vegetables. Though there are also many products one would find in a supermarket, the ones I mention have an application in their society of origin, a different culture.

In most of the developing world the average person (the vast majority of people are very poor by northern, "developed" world standards) eats the same basic grain every day, often for every meal. Most of the foods mentioned above are basic for some culture or society. For instance, North Americans normally encounter millet on a regular basis as tiny, golden balls in bird seed. Yet it is the staple food of a large number of Africans. Discussing it, the Finnish nutritionist Paavo Airola has written:

> Millet is a truly wonderful, complete food. It can rightfully be called the king of all cereals possibly sharing this distinction with buckwheat. It is high in complete proteins and low in starches. It is very easily digested and never causes gas and fermentation in the stomach. . . . Dr. Harvey Kellogg, famed nutritionist, said that millet is the only cereal that can sustain or support human life when used as the *sole item in the diet*. Besides complete proteins, millet is rich in vitamins, minerals and important trace elements, such as molybdenum and lecithin. (Airola 1974, 249).

Dried split yellow peas and other split legumes are the basis of East Indian spiced dal. In *The Vegetarian Epicure*, Anna Thomas has written: "This is one of the most highly concentrated forms of protein, and a staple food of India—a country largely vegetarian for many reasons. . . . It is simple to prepare, ridiculously inexpensive and delicious" (1972, 263).

Similarly, bulgar and couscous are grains of the Middle East. Soy, in milk, or tofu is the basic grain of Asia. But because soy is a complete protein (see chapter 2), it can be used by the one-fourth of the world's people who do not produce (as adults) the enzyme lactase that allows them to digest milk. Most Africans and many Asians fit this category. During the early days of food aid, milk shipped to Africa often caused such indigestion that people used it for whitewash.

The stone-ground, organic hard and soft wheat flour, the organic raisins and dates and even noninstant milk are "relics" from our past. Because steel rollers for grinding wheat and petroleum-based chemicals in agriculture are of relatively recent origin, that part of humankind that ate wheat (not everyone did) ate it in these forms. *All* fruit, grains, and vegetables were organic. Pesticides, fungicides, herbicides, and petroleum-based fertilizers were not invented until this century. During, and prior to, the Industrial Revolution, the basic diet of the English working class was beer and wheat (and oat) bread, often spread with some marmalade—a source of vitamin C.

In the Eastown Food Co-op there are organic manufactured products as well. These include frozen dinner items such as vegetarian egg rolls, okra patties, split pea and lentil soups, pot pies, stuffed manicotti, spinach lasagna, and tofu patties (with no cholesterol).

Before their farm burned, organic farmers supplied organic chicken to the co-op. Now they offer free-range chicken with "no hormones, growth promoters or antibiotics" (according to their sign). Now the chicken at least exercises. As a result, chickens are significantly larger, better tasting, and without the yellow fatty deposits in the muscles that characterize commercial poultry.

Newer organic convenience foods are obviously directed to fulfill the needs of a constituency whose lifestyle may need faster meals but whose food consciousness guides them away from the chemical pharmacopeia contained in a good many prepared products of mainstream American supermarkets. These people seek the convenience of foods that can easily be heated and eaten. But their food consciousness dictates that they do not want to subject themselves to the risks that many of the preservatives, coloring agents, flavor and taste enhancers, stabilizers, and sweeteners entail. This constituency is large enough in the United States and England that many supermarkets have added organic products. But there does not yet seem a large enough constituency for the chain stores to stock the great variety of organic products that health food stores and food co-ops do.

Rice

Of the staple grains with which Americans are familiar, rice stands as the most famous. Growing up in the 1950s, my upper-middle-class WASP mother prepared instant short-grain white rice smothered with *butter* and *sugar*! Like most upper-middle-class families in the 1950s, my family knew little about nutrition and health.

In much of the rest of the world, rice is *the* important grain. In much of Asia and Latin and Central America a variety of other grains is often available, but eating rice is the indication of nonpoverty. Wheat, which is less productive per land area, has become a status food (replacing millet and sorghum) in much of Africa and is becoming a status food of Asia.

The Eastown Food Co-op carries six varieties of rice: five are organic, one is not. The five organics are long, medium-, and short-grained brown rice, and white and brown basmati rice. The co-op stocks one long-grained, nonorganic white rice. There is no short-grained instant white rice in the bins.

Compared to wheat and barley, rice is much more productive under most conditions. But where it is adopted, there are major implications for the society. In a sense, the personality characteristics of sensitivity, cooperation, persistence, and flexibility (needed to grow rice) often describe Asian rice-based cultures. They are significantly different from the coercive, controlling ones that accompany Western forms of food (Visser 1986, 162).

Other implications of rice for society include its influence on health, political structure, social class, language, sexual divisions, spirituality, mythology, and manners. Because rice has no cholesterol, it is usual for Asians to survive the Western artery and heart diseases. The Japanese now have the longest life spans in the world.

Rice (usually) requires large quantities of water, delivered in a precise manner. Therefore, a people who adopt it also commit themselves to large-scale administrative units. Water must be found, transported (often via massive, publicly created aqueducts), pumped, and sluiced. All these activities require centralized organization ruling over coerced peasant masses (Jones 1989, 9). This political concentration was foreign to the yeoman farmers of the United States and the feudal lords of medieval Europe.

Rice has often been an indicator of social distinctions. As Visser states: "The rich gourmet was endlessly finicky about rice types, provenances, aromas and cooking styles, the poor had to be content with brown rice or gruel (congee)" (1986, 165). Ever since milling was discovered, white rice has generally gained high status. Yet it lacks significant B vitamins. In India, milled rice maintains most of these vitamins because it is parboiled.

In addition, rice relates to the sexual division of labor both in myth and contemporary society. Malaysian mythology cites a mythological "fall" of humanity from a violation of rice rituals (Visser 1986, 169). In addition, "rice . . . is everywhere a woman. She is tender, beautiful, and timid, and she dislikes being handled by men. The inviolability of sex roles is heavily underlined by the preferences of rice. . . ." (Visser 1986, 166).

Rice is even reflected in metaphors of the language. In China, "a southerner who has not eaten rice all day will deny having eaten at all, although he or she may have consumed a large quantity of snacks" (Anderson 1988, 139). And "an 'iron rice bowl' is the modern Chinese expression for job security, while being unemployed is 'breaking the rice bowl'" (Visser 1986, 164).

Modes of Ownership

Like rice, each food at the Eastown Food Co-op has a social history and current use. Each has growing and production methods that have implications for the social structure in which they are produced and eaten. In essence, the Eastown Food Co-op is a culinary library of many of the societies and cultures of the world. By learning about the production, methods of preparation, and methods of consumption of these foods, one gains an education into the increasingly diverse world in which we live.

Why is the co-op so different from the "standard" grocery stores? The answer entails a thorough analysis of the different modes of ownership that structure each organization. The Eastown Food Co-op, like most food co-ops, is a democracy. It is governed by a board of directors elected from the membership (membership consists of paying a buying deposit and undergoing an orientation process). The board of directors hires the manager. The manager hires the staff (many co-ops still use all or part of their membership as part-time, volunteer staff).

Because the co-op is a democracy, there is broad discussion of most major issues. Recently the discussion centered on whether to go public (the system of membership deposits is a private system). The March 1991 edition of the co-op newspaper, *The Bean Sprout*, listed the arguments for and against going public. And it stated, "You are encouraged to address this issue *in writing.*"

As a result of the democratic governance structure, such other issues as the ethical sources of products (whether it comes from an exploitative source) or the chemical products used (is DDT used in a developing world country?) are addressed. The co-op often buys from food-producing co-ops. They purchase some coffee from a Nicaraguan co-op.

Periodically, prices of items are compared with those at the standard grocery stores. The co-op does not always come out lower. In spite of not having to put its profits into the pockets of multinational corporations or stockholders, the economies of scale and oligopoly on which larger stores operate make it difficult to compete on the basis of price alone.

Oligopolist Ownership Structures

Differences between co-ops and grocery stores or supermarkets in much of the world lie in the differences between the structures of ownership of food co-ops and other food stores. The first element of which most shoppers are aware is that the vast majority of supermarkets are parts of "chains." These chains are often a small part of a corporation that is a vertically integrated oligopoly.

What we tend to think of, and are encouraged to think of, as agriculture (the realm of the family farm), food production (for the most part), and food selling have been the latest major areas of the American economy to be reorganized along oligopolistic patterns. The decline of the family farm is discussed in chapter 8. Here we note how oligopolization affected the selling of food products to the public—the supermarket.

Whereas the Eastown Food Co-op is run democratically for the benefit of its membership, oligopolies are run for the maximization of stockholders' profit. In order to do this, it is desirable to minimize risk and uncertainty in the system. While state socialist systems were characterized by state planning, oligopolies find it to their advantage to decrease uncertainty by owning their growers, harvesters, processors, manufacturers, and other suppliers. This process began with the Heinz family's ownership of all the elements from farm to grocery suppliers in the 1890s (Levenstein 1988, 36-37).

Box 1.1 OLIGOPOLY

To explain oligopoly, one must briefly survey the history of the American (and world) economy in the past century. Most Americans think of our economic system as one characterized by free enterprise and the pursuit of profit.

This Adam Smith version of entrepreneurial capitalism was true at the beginning of our economic system. In fact, in the previous century, food stores competed against each other on the basis of *quality* and *price*. Customers had (within the limits of their geographic mobility) a choice of stores from which to purchase their food. They could freely choose among them. This was the situation for most of American industry before 1900.

In an economic system where units compete against each other on the basis of efficiency (much like the Darwinian version of evolution), generally the most efficient survive and the less efficient—and less profitable—die out.

This situation began to change in such industries as automobiles after the 1929 depression. That economic phenomenon accelerated the process of the elimination of the weaker companies and speeded the growth of the stronger ones. Antique car shows now demonstrate the early variety of American automobile production. Fuller Touring cars, Jacksons, Argos, Brisoe, Cyclops, Earls, Hudsons, Cord, DeSoto, Surry, Duesenberg, Beuz, Tucker, Crosleg, Flint, Falcon, Ajax, Pilot, Meteor, Stutz Bearcats, LaSales, Jaxon Steam cars, Nash Ramblers, Studebakers and many others did not survive the process of competition through quality and price.

In the search for greater profit, Henry Ford's assembly line for the production of Model A and Model T Fords spelled the death of most individually, hand-produced American cars for the general market.

By 1950, there were but five car companies in the United States. And three, General Motors, Ford and Chrysler controlled the vast majority of the business. When fewer than four companies control more than 50% of an industry, economists refer to that mode of industrial ownership as an *oligopoly*.

In an *oligopoly*, industry operates very differently from entrepreneurial capitalism. Because it is no longer in the interest of any of the oligopolists to compete through quality and price, they engage in "price leading." When one raises prices the others approximately follow suit. In an oligopoly, it is in *all* their interest to do so. Thus they benefit at the expense of the consumer. And, unlike entrepreneurial capitalist enterprises, which can *drop* the price with improved technology, generally prices in an oligopoly go in one direction only—up.

In addition, the effect of this kind of ownership on technology is significantly different. Rather than competing through *real* technological advances, competition is done through *advertising* and *style*. The major mechanism of competition in the 1950s, the United States' first full-fledged era of oligopoly capitalism in the automobile industry, was the introduction of such style changes as two-tone cars and fins. Even though the technological improvements, which could have made cars safer, more efficient, and longer lasting, were known, they were not used. Such technologies as radial tires, disc breaks, hypopneumatic suspensions, the stratified charge engine, and seat belts awaited competition from another country—Japan—before American car companies were forced to introduce them. Pre-Japan oligopolies in America kept important, efficient, lifesaving technologies from the American automobile because they thought it was profitable to do so.

That process is termed *vertical integration.* It is complete when one company owns every part of the production process from farm to supermarket chains, as with the poultry industry. It is a mode of ownership that now characterizes most of the food produced and sold in the United States.

In agriculture, over 80 percent of family farms have succumbed to competition with oligopolistic farming. The first stage was "contract farming." Here the farmer agreed to purchase from and sell back to the dominant company. Even seed and fertilizer are now often controlled by the oligopoly. As Ann Buchanan writes of the oligopolistic situation: "At the farm end, four firms control 68% of the petroleum products, 74% of the agricultural chemicals and 80% of the rail transport. Just two firms supply half the hybrid seeds" (1982, 86). And "at the other end of the process, Susan George tells us that over 55 percent of the market is controlled by four or fewer firms (the definition of 'oligopoly' conditions) for **every** major food category in the United States; and in many categories the percentage is far higher" (George 1979, 27).

Just like the automobile industry, companies producing and marketing food compete through advertising and product proliferation because the key is profit (not health or savings, as with the democratic co-op) instead of competing through price and quality.

Oligopoly in Breakfast Cereal

Oligopoly is well researched in cereals. In February 1981 *Consumer Reports* added a supplement to its report on them entitled "Monopoly on the Cereal Shelves?" It evaluated the economic state of the market:

> There is ". . . little or no price competition in the cereal industry, according to the FTC [Federal Trade Commission]. The $2 billion industry is

> what the FTC calls a 'shared monopoly.' Four firms: Kellogg, General Mills, General Foods and Quaker Oats control nearly 85% of the market. . . . They earn above-normal profits, exclude new competitors, and overcharge customers. The overcharge for cereals amounts to more than $100 million a year ... and results from prices set higher than they would be if the market were competitive." (p.76)

These companies match their price increases so well (with Kellogg as the "price leader") that, as revealed by the FTC, "from 1958 to 1970 the industry's rate of return on capital . . . was 19.8% . . . double the average rate of return for all manufacturing industries" ("Monopoly on the Cereal Shelves" 1981, 77). These profits exceed all manufactures except drug companies (DeLind 1990, 7).

The manner in which companies, still motivated by profit, "compete" in an oligopoly (or shared monopoly) is through *product proliferation* (the analog to automotive "style"). Columbia University nutritionist and food authority Joan Gussow notes that the food industry introduces from nine to thirty-three products *per day* (1991, 6). *Consumer Reports* writes:

> Take a plain, puffed-corn cereal such as *Kix*. Add a fruit flavor, and you have a "new" product called *Trix*. Make the flavor chocolate, and there's yet another "new" product, *Cocoa Puffs*. Are vitamins all the rage? Spray those same corn puffs with vitamins and minerals and you have *Body Buddies*. Even *Body Buddies*, the great grandchild of *Kix*, has begotten two children of its own—*Body Buddies* with brown sugar and honey, and *Body Buddies* with natural fruit flavor ("Monopoly on the Cereal Shelves" 1981, 77).

These companies claim that they are "only giving the consumer what she or he wants." Is that true? In the context of an oligopolistically controlled market, consumer "choice" is something much different from the consumer driven market of competitive or entrepreneurial capitalism. Instead of an intelligible amount of information about products on which consumers could make an informed choice, the consuming public is flooded with a "'fire hose of information' coming from everywhere about food, nutrition and health" (Gussow 1991, 3–4). Much of that information is an added attempt to delude the customer about the true contents of the food. The oligopolists in the food industry, motivated by collective profit, have been resistant to simplified, understandable food labeling (Michael Jacobson, CPSI Research). As Joan Gussow writes, "the advertising about these products is designed to sell them, not to educate about them" (Gussow 1991, 6).

The food industry uses survey information to conceptualize consumer interest as convenience, quality, variety, excitement, nutrition, safety and health (DeLind 1990, 11, quoting Oppedal 1989). But, as DeLind writes: "It can hardly be coincidental that each of these consumer sub-values is also consistent with the interests of agribusiness" (1990, 12). The multitude of daily new products fit the profit needs of these vertically integrated corporations. Furthermore, those "values that apparently drive consumer behavior are reinforced by a multi-billion dollar campaign of advertising designed to tell consumers what it is they really want" (DeLind 1990, 14). "Demand" is largely created.

Part of this dynamic is the deceitful manner in which products are labeled. The newsletter of the Center for Science in the Public Interest (Washington, D.C.) has a monthly listing of "Food Porn." The following are some examples of the kind of deceitful advertising and "information" that characterize food labelling in regular grocery stores.

Good Coffee? Fat Chance!

Here's a recipe for a celebrate-the-moment cup of coffee.

Take a teaspoon of instant coffee, add four teaspoons of powdered coffee whitener, toss on a teaspoon of table sugar, add hot water, and stir.

Don't want to go to all that trouble? Well, don't worry. The wonderful folks at *General Foods* have done it for you—in their pre-mixed *International Coffees*. . . .

. . . you'll notice that you're paying $6 a pound for something that is, at most, 25% coffee.

And . . . there's more hydrogenated coconut oil than any other ingredient in International Coffee. Drink two cups and you'll deposit six grams of the artery-clogging fat in your system . . . (*Nutrition Action Health Letter*. March 1988, 16).

From the look of Del Monte's brilliant pink *Strawberry Yogurt Raisins*, you'd expect a snack that far surpasses ordinary dried fruit. Why eat plain raisins ... when Del Monte is offering the vitamin C and fiber of strawberries, and the calcium of yogurt?

But, a close look at the side label could change your mind. The principle ingredient in Strawberry Yogurt Raisins isn't strawberries or yogurt or even raisins. It's the "strawberry flavored yogurt coating," made largely of sugar and hydrogenated coconut, cottonseed, palm, palm kernel or soybean oils . . . .

Worst of all, 38% of the calories in Strawberry Yogurt Raisins come from fat. Who wants raisins wrapped in fat—a full teaspoon in every serving—

> when regular, uncoated raisins are fat free? . . . (*Nutrition Action Health Letter*. March 1988, 16).

In contrast, at the Eastown Food Co-op there are plain plastic bags of "commercial" and "organic" raisins!

> The new *Milky Way Shake* is now available in the freezer case of your local supermarket. Each 10.5 ounce container is packed with twice as much fat as you'd get in a cup of whole milk . . . .
>
> The shake gets 40% of its 380 calories from fat. Compare that to the wimps at Burger King and Arby's (20% to 30%) or Hardee's and Jack-in-the-Box (15% to 20%). And the shakes at McDonald's, the industry leader, get only 5% of their calories from fat.
>
> Its not hard to see what sets the Milky Way apart from its competitors. It isn't bulking agents like cellulose gel and parageenan. Everybody uses them. Its the whole milk and grade A cream—the stuff that blocked blood vessels are made of . . . (*Nutrition Action Health Letter*. January-February 1991, 16).

From these examples, one gets some insight into the radical differences between a standard supermarket/grocery store and a democratically run food co-op. The latter has none of the hoopla, deceitful "informational" advertising, and excessive packaging that characterize the supermarket. If one wants a granola, one goes to the plexiglass bins of granola and reads the plainly typed ingredients. There is no attempt at camouflage or "sell."

The result is that contemporary supermarkets, run in the interests of oligopolies (whose primary motivation is profit), are culinary minefields of bad nutrition. Joan Gussow (1990) says the best advice she can give shoppers is to "shop the periphery of the store," where the fruits, vegetables, meat and dairy products are usually kept.

This situation is made worse by the truly massive number of products in modern supermarkets. Their number is justified by the notion of "freedom of choice." But Gussow writes:

> . . . the notion of *free* choice usually also implies *informed* (free) choices. And informativeness and variety are to a large extent inversely related. The greater the number of products from which the consumer must choose, the less time he or she has to be knowledgeable about any one of them (choosing wisely from among 60 yogurts is undeniably more difficult and time consuming than choosing wisely among five). So the greater the quantitative choice in the marketplace, the less able the consumer becomes to make careful discriminations. (1990, 8)

The resulting situation, without believing too much in conspiracy theories of marketing, is one in which food producers know (they employ some rather high priced social scientists to help them) that more varieties cause *less* informed choice. Product proliferation in an oligopolistic situation is a superb solution to reducing the effectiveness of nutrition education. By confusing or overloading the informational resources of the consumer, companies encourage irrational rather than reasoned choice. Unless one has a veritable food computer mind, she or he would do better to at least shop where there is a modicum of trust between food seller and food buyer.

At one time, when food shopping consisted of a personal relationship between a butcher, a fruit and vegetable stand, a grain and other specialty stores, one could have some feeling that the owner was looking out for the consumer's interests. In the era of mass food merchandising, the check-out clerks may smile and ask "How are you doing?" But the socioeconomic system behind their products makes it all but impossible to protect oneself from the technologies of advertising and mindless product proliferation/pollution.

One solution is to shop where one elects the board that hires a responsible manager. When was the last time a democratic supermarket fired the manager for irresponsible food selection? A food co-op provides a mechanism for shoppers' participation in their food destiny. This is the foundation of an authentic democracy.

Summary

This chapter has analyzed chicken as an example of the transition from pre– to industrial production, processing, and consumption. We have seen the application of industrial technology applied in the context of oligopolistic capitalist profit seeking. In the process, the quality of the food, the workers, and the general population have deteriorated.

In contrast to this form of ownership and technology, the Eastown Food Co-op tries to preserve the good elements of past food cultures and to adopt contemporary technologies that actually improve the quality of life and eating. Co-op frozen fast food is nutritionally healthful. Furthermore, the co-op preserves an atmosphere of humane relationships. Through its democratic organization and its intimate layout, people actually talk to each other as part owners. You can ask why "we" don't carry any specific item, rather than why "you" don't carry it.

The co-op serves as a living museum of food from around the world. One can learn about various parts of the world by preparing and eating the food of various cultures. And one can understand much about food from attempting to become more involved with issues about food growth and processing.

Decisions to buy as much organic produce as possible are a function of the democratic governance structure.

Furthermore, over a number of years the co-op has conducted an ongoing discussion of the morality of carrying various products. Should we carry table grapes from California that the farmworkers have been boycotting? Should we carry coffee from El Salvador where workers are blatantly exploited? Or should we intentionally purchase from producer cooperatives and socialistically oriented societies? All these questions have been repeatedly asked. And the answers have sometimes changed with different elected board members.

The food co-op movement is itself a challenge to the marketing of mainstream American food. And no one knows if it can survive the copying that some chain stores are doing in the areas of organic food and open-bin portioning.

But before systematically exploring the domain of contemporary American food systems, it is necessary to look at the history of food production, distribution, and consumption. This is chapter 2.

REFERENCES

Airola, Paavo. 1974. *How To Get Well.* Phoenix: Health Plus Publishers.

Anderson, E.N. 1988. *The Food of China.* New Haven: Yale University Press.

"Baby Formula and The Big Boys." 1991. *The Grand Rapids Press*, January, A10.

Bates, Eric and Bob Hall. 1989. "Ruling the Roost." *Southern Exposure*, Summer, 11.

The Bean Sprout Flyer. 1991. Grand Rapids, Mich.: Eastown Food Co-op, March.

"Clean Arteries Fixed." 1991. *Nutrition Action Health Letter*, January/February, 16.

DeLind, Laura. 1990. "The Well Managed Consumer: A Notion of Convenience and Value for Michigan's Food and Agriculture Industry" Unpublished paper at Michigan State University. Lansing, Michigan.

Devine, Tom. 1989. "The Fox Guarding Her House." *Southern Exposure*, Summer, 41–42.
George, Susan. 1979. *Feeding The Corporate Few*. Washington, D.C.: Institute for Policy Studies.
Goldofas, Barbara. 1989. "Inside the Slaughterhouse." *Southern Exposure*, Summer, 27–30.
"Government Probes Formula Pricing." 1991. *Consumer Reports*, March, 142.
Guess Who's Coming to Breakfast? n.d. Stoughton, Mass.: Packard Manse.
Gussow, Joan Dye. 1991. *Chicken Little, Tomato Sauce, and Agriculture*. New York: Bootstrap Press.
Hall, Bob. 1989. "Chicken Empires." *Southern Exposure*, Summer, 12–17.
Jacobson, Michael. 1992. Letter, 1–4.
Jones, E.L. 1987. *The European Miracle*, 2nd ed. New York: Cambridge University Press.
Levenstein, Harry A. 1988. *Revolution at the Table: The Transformation of the American Diet*. New York: Oxford University Press.
Lentz, Charles M. (1988. "Grocery Shopping in the 1930's." *Michigan History*, March/April, 14–15.
"Milky Way Shake." 1991. *Nutrition Action Health Letter*, January/February, 16.
"Monopoly on the Cereal Shelves." 1981. *Consumer Reports*, February, 76–80.
Oppedal, A. 1989. "Defanging Anti-Trust." *Hawthorne Management*, 14.
Pohl, Frederick, and C.M. Kornbluth. 1985. *The Space Merchants*. New York: St. Martin's Press.
"R & D at Kellogg's." 1987. *Nutrition Action Health Letter*, March, 16.
"Strawberry Yogurt Raisins." 1988. *Nutrition Action Health Letter*, March, 16.
Tannahill, Reay. 1973. *Food in History*. New York: Stein and Day.
Thomas, Anna. 1972. *The Vegetarian Epicure*. New York: Vintage Books.
"USDA Urges Irradiation of Poultry to Boost Safety." 1992. *Grand Rapids Press*, May 7, A3.
Visser, Margaret. 1986. *Much Depends on Dinner*. New York: Grove Press.
"Zapping Irradiation Away." 1992. *Nutrition Action Health Letter*, March, 16.

Chapter 2

Food and Nutrition in Pre-industrial Societies

Food and Nutrition in Hunting and Gathering Societies, 12000 to 7000 B.C.

We know about the earliest history of human eating from fossil records and contemporary hunting and gathering groups. Neither method of study is ideal. Contemporary hunters and gatherers often scavenge on now relatively denuded land. We must assume this was not the case for the hunters and gatherers who appeared 14,000 years ago and predominated until about 7000 B.C.

There are alternative anthropological interpretations of the roles of men and women in hunting and gathering societies. Until the past two decades, the reigning hermeneutic about hunting and gathering societies was that men did the hunting and women did the gathering. Hunting was judged more important. It generated the necessity of coordination, language and tools.

A popular Harvard University film, *The Hunters*, shows a contemporary tribe, the Kung Bushman, as they hunt and gather in the Kalahari Desert of southern Africa. The film devotes over 80 percent of its footage to the hunting activities of the men. Made in the 1950s, probably by men, we find an androcentric (male centered) perspective informing the shooting and editing. But new perspectives and new data have changed some of our evaluation of the relative importance of the roles of men and women in hunting and gathering societies. Grashuis calls our attention to Elise Boulding's position that, in nomadic societies, 80 percent of the food was gathered by women (1989). Similarly, Frances Dahlberg's *Woman the Gatherer* contains numerous recent contributions challenging the previously accepted versions of prehistory.

Dahlberg cites evidence that "male archaeologists had not collected, described and catalogued large numbers of the grinding stones that women used" (Dahlberg 1981, citing Kraybill 1977). She notes that, in all probability, the abundance and predictability of gathered food made hunting possible. And she mentions that language might as realistically have arisen as the result of shared activities between mother and child as in hunting. Surely there is little evidence for the popular notion that hunters need to talk while planning and executing the hunt, "since wolves, chimpanzees, and modern hunters are silent while hunting" (Hockett and Ascher 1964, cited in Dahlberg 1981, 10).

Modern use of the *Ethnographic Atlas* shows that in fifty-eight foraging (hunting and gathering) societies, about one third of the food comes from hunting and two-thirds from gathering by women. With these feminist hermeneutics, the relative importance of men and women is reversed. Zihlman states that "gathering and not hunting was the initial food-getting behavior that distinguished apes from humans. It was an innovation whereby human females used tools to obtain food" (Zihlman 1987, 93, cited in Dahlberg 1981).

Nonandrocentric anthropologists have also begun to observe tribes other than ones in which men do all the hunting. Agnes Estioko-Griffin and P. Bion Griffin conducted a study of the Agta tribe in the Philippines. There "women participate in all the subsistence activities that men do" (1981, 126, cited in Dahlberg 1981). And they noted that "the disadvantage of the early division of labor may be considered. In acquisition of small game, all but the most pregnant females should be competent" (cited in Dahlberg 1981, 145).

These interpretational divisions are important for our knowledge of nutrition insofar as they bear on the question of what human beings evolved to. Because biological adaptation is so slow, one can strongly argue that we still have 14,000-year-old bodies in the modern age. If we know what humans ate originally, we should have some idea of what we should be eating now.

According to the androcentric interpretation, human beings were made to eat meat. H. Leon Abrams, Jr., argues that, with man the hunter, "from the very beginning the diet of humans has been meat orientated" (1979, 39). And he argues, contrary to most contemporary nutritional science, that, within any given society, "those who develop coronary heart disease do not necessarily eat differently from those who do not" (Abrams 1979, 40). This parallels nicely the argument that Marvin Harris (1985) makes that meat is the most economical form of protein and therefore the most useful for human beings.

Unfortunately the argument does not stand up when we look to *contemporary* meat eating as the best nutrition. Historically the meat quality has changed. The meat that hunters and gatherers ate is what we would now term *organic*. There were no human-created chemicals in it. And there is evidence that it had more good cholesterol (high HDL) because of the "exercise" animals normally do. It was not marbled meat. It was lean, muscular, and organic.

In addition, one who accepts recent feminist-oriented interpretations of the importance of gathering should raise some questions about Abrams and those who postulate a "garden of Eden" of carnivores (meat eaters). Although meat was part of people's diets, it seems that it was closer to the Chinese pattern of eating a small amount of meat mixed with a basically vegetarian cuisine than to that of the 1950's American steak and hamburger culture.[1]

Though this review of contrasting interpretations of prehistory may seem a long detour, the reader can now appreciate how these facts inform nutritional knowledge. In addition to the question of ratios between meat and grown foods, one can also make some suppositions about the frequency of eating. In all probability, hunters and gatherers neither sat down to three family meals a day nor went to the prehistorical equivalent of McDonald's for their ration of junk food. People ate sporadically, but sometimes almost continually. This is much like the pattern recommended for hypoglycemics and diabetics in our society. Sometimes we call it "grazing." It may have been our original, natural pattern.

Foods were mostly fresh (there was no refrigeration). There were no preservatives except salting and drying. Whatever could be eaten was probably eaten raw. But fire improved the palatability of food and retarded spoilage. It improved the digestibility and taste of some grains and meat.

Finally, both groups of anthropological interpreters agree that sharing (social "refrigeration") was an effective adaptive mechanism. Hunting and gathering societies were mostly egalitarian. By spreading the fruits of their labor among those who may have been less successful in hunting and/or gathering on any given day, they ensured the success of the entire group's survival. Could it be that sharing is also a "natural" human trait?

Food and Nutrition in Food-Growing Societies, 7000 B.C. to A.D. 1850

The origin of food growing societies is termed the Neolithic Revolution. There is some dispute about the origin of growing food. Whereas it has previously been assumed that women discovered that grain could be grown and took advantage of that technology, many scholars now assume that the technology was known before it was used. Only population pressure generated by increased numbers and by the overhunting of game provided the motivation to move from the somewhat idyllic life of hunters and gatherers to the more labor-intensive life of farmers. While food was fairly equitably distributed in hunting and gathering societies, the growing of a basic food crop culminates in a highly inegalitarian society and therefore a highly inegalitarian distribution of food.

Preindustrial food growing societies developed in two stages. Horticultural society—the growing of gardens—dominated from about 7000 B.C. to 3000 B.C. And agrarian or agricultural society dominated from 3,000 B.C. to about A.D. 1850.

The basic economic change caused by the evolution of food growing was the generation of an economic surplus. When there was more than enough food (at least for some), the extra could be consumed (conspicuous consumption leading to obesity), wasted (*potlatches,* warfare), traded (for further riches), or used to support occupations not directly dependent on food growing. While the horticultural surplus was generally minimal, that of many later agrarian (agricultural) societies was reasonably large. The Roman and Chinese empires were part of extensive agrarian societies. Here an elite of 2 to 10 percent of the population lived extremely well at the expense of the rest. Inequality was the most extreme in the history of humanity.

Horticultural Societies 7000–3000 B.C.

Agriculture emerged independently in at least five areas of the world, which are termed the Vavilov centers of diversity: the Middle East, China, Southeast Asia, Meso America, and Peru (Sanderson 1991, 72). In the Middle East, wheat and barley were the main crops. In China and Southeast Asia, millet (see chapter 1) emerged first, followed by rice, yams, and taro. "The most significant New World plant domesticate was maize, the wild ancestor of which was a plant known as *teosinte*" (Sanderson 1991, 72). Other plants from the New World included "amaranth, quinoa, lima beans, squash, tomatoes, potatoes, chili peppers, and cacao" (Sanderson 1991, 72).

Techniques of food growing probably evolved slowly. At first, wild grains were merely gathered. This allowed the development of ways to store, grind, and cook the grains. As population pressure increased, techniques of food growing evolved. As Sanderson writes:

> Most simple horticulturists live in heavily forested environments and practice a form of cultivation known as *slash and burn*. . . . This cultivation technique involves cutting down a section of the forest growth and then setting fire to the accumulated debris. The ashes that remain serve as a fertilizer, and usually no other fertilizer is added. The crops are then planted in these cleared plots (usually no more than an acre in size) with the aid of a digging stick, a long pole with a sharpened and fire-hardened end. A given plot may be devoted to a single crop, but a more common practice is to plant several minor crops along with one main staple. (1991, citing Sahlins 1968).

The task of clearing and preparing the plots generally falls to the men, while that of planting and harvesting is typically considered the responsibility of women (Sanderson 1991, 73).

Horticulture allowed more stable and larger societies. Whereas hunting and gathering societies usually numbered 30 to 60 people, horticultural ones could range to 150. Because these latter societies were not always moving, they could build small settlements and develop such technologies as pottery. Eventually, through experience with heat, pottery technology led to metallurgy.

The process of farming generated a small economic surplus. Generally these tribes had some kind of chief or leader and a few retainers. Nevertheless, the *degree* of inequality was small relative to that of the later agrarian society.

Often war, as portrayed in the anthropological film *Dead Birds*, occupied periods between work for the men. Because they had relegated the routine farming tasks to women, the men now had the leisure to indulge in a (from an American perspective) horticultural equivalent of football. This war/sport also served as a form of population control through occasional deaths.

Because of woman's importance in food production, they were accorded a higher status in horticultural societies. The Lenskis note that in 26 percent of known horticultural societies, one traced one's lineage through the mother (1982, 154). This approximate percentage continued through advanced horticultural society.

The technological improvement that distinguished advanced horticultural societies from simple ones was metallurgy. Metallurgy made possible the hoe in agriculture and, eventually, bronze weapons in China. In horticulture, the hoe made it possible to dig a bit deeper, destroy weeds, and generally use the garden plot more intensively. It is commonly thought that in this era, because of the more intense use of the land, manure and humus fertilizers were employed. This practice generated a larger crop and therefore a larger economic surplus. As a result, settlements became larger, and the size and degree of inequality in these societies increased.

In China, where bronze replaced copper in weapons of war (but not in hoes), a small warrior nobility dominated a large mass of peasants. It now became profitable to conquer people (as well as nature), for the captives could be employed as slaves in producing even bigger crops. This large amount of labor allowed the development of the intensive, coordinated agriculture necessary for irrigation in a rice-growing society.

Throughout horticultural society, the amount of work required to acquire food intensified. Nutritional standards generally declined for all but the elite or governing class. There was a shift from a diet with adequate (if sporadic) meat (if the tribe was successful) to one in which there was a much larger percentage of grain—often the same grain. As I discuss in chapter 3, a vegetarian diet is quite capable of sustaining life. But, with most grains, the amino acid balance (protein) must be corrected with *complementary grains* to make complete proteins. If this is not done, protein deficiencies can develop, leading to disease and death.

The anthropological sites that have yielded useful data about horticultural and agrarian societies are probably the ones in which people accidentally happened on the correct mix of complementary grains or in which they supplemented their grain intake with meat (either domesticated or wild) or milk (both complete proteins). Of the societies that did not happen on this knowledge, we know little because we may suppose they did not survive long enough to develop viable civilizations whose remains could be studied.

Agrarian Societies: 3000 B.C.–A.D. 1850

The transition to plow technology distinguishes agrarian (from *ager*, "field") from horticultural (from *hortus*, "garden") societies. The plow solved the problems of weeds and nutrients in farming. Through its greater penetration, it allowed the turning over of the deeper nutrients from the soil. It also destroyed the weeds, loosened the soil, and plowed the humus back into the soil. Altogether, the plow made for an enormous increase in food production.

Sanderson quotes Boserup as citing population pressure as the motivating factor in the technological development of the plow (Sanderson 1991, 82). This interpretation assumes that people work only as hard as is necessary in order to have reasonable amounts of food. When the population pressure built up, farmers began experimenting with the plow technology.

Initially the plow was a large hoe pulled by humans. With experience it was reshaped to pull through the soil more easily and, eventually, to be pulled by animals. This made possible the cultivation of a significantly larger area by fewer people.

The plow stands in a progression from hunting and gathering society, where labor becomes more onerous. Whereas hunting and gathering societies had a fair amount of leisure (where food was plentiful), horticultural societies involved the practice of breaking ground by the men and planting, care, and harvest usually by the women and children. Agrarian society required significantly larger amounts of labor. Just as family farming today requires labor from sunup to sundown, peasants were kept significantly busier than in the previous form of society.

In spite of the greater amount of labor required, agrarian societies generated a significantly greater product as a result of technological advances. They had a much larger economic surplus, from which developed most of what we know as "civilization." Because the laborers of this society could support a larger class of people who did other things (such as reading and writing), we have the development of the accouterments of civilization.

Just as the initial surplus of horticultural society allowed a small degree of social differentiation and stratification, the significantly larger surplus of agrarian society allowed a larger and more highly stratified (of greater social distance) elite. Generally the elites of agrarian societies constituted between 2 and 10 percent of the population.

Religion and coercion served to extract the economic surplus from the working population. In Egypt, Pharaoh was god; sacrifices and contributions were due him. The Roman Empire combined the carrot of religion with the stick of the tax collector. Generally, the tax collector was given the authority to extract as much as he could from the peasant population of a region.

Agrarian societies divided by the material from which the plow was made. Earlier agrarian societies manufactured the plow from wood. Metal was usually reserved for the elite. It was used for weapons and jewelry. But there are recorded instances in China of adding brass to the blade of the plow to make it cut deeper and last longer (Lenski and Lenski 1982, 170).

Advanced agrarian societies are undergirded by the iron plow. Iron was originally significantly more valuable than gold. Because peasant labor was cheap, iron was again used first in weapons and jewelry. Perhaps with the pressure of increased population, it became obvious that iron could be used to make better plows. The Lenskis credit the Hittites of Asia Minor with the discovery of iron ore and smelting techniques (1982, 180). At the collapse of their empire in the eighth century B.C., the technology of smelting spread throughout the known world.

Just as food consumption followed the degree of stratification of horticultural society, so in agrarian society there were significant differences in food consumption. But because of the greater social distance in agrarian society, there was significantly greater inequality in quality and quantity of the food consumed. Two examples of this pattern are India and Western Europe. In India ". . . the caste system itself is partly defined in terms of the type of food a man is allowed to eat" (Goody 1984, 115). There are prohibitions on both intercaste marriages and intercaste dining (Goody 1984, 114).

Caste distinctions are maintained by both quantity and quality of food. "The meal of a gentleman . . . consisted of one *prastha* of pure unbroken rice, one-fourth of a *prastha* of pulses, one sixty-fourth part of a *prastha* of salt and one-sixteenth part of a *prastha* of clarified butter or oil" (Goody 1984, 115). "For menial servants the quantity of pulses prescribed is one-sixth of a *prastha* and the quantity of oil or clarified butter half that prescribed for a gentleman" (Goody 1984, 115). Millet plays a greater part in lower-class diets, as rice does in the upper-class ones (Goody 1984, 125).

In India (as well as Western Europe) it was the luxury of the rich that they could fast. The highest spiritual status was expressed by vegetarianism and abstinence. But one must realize that "abstention only exists in the wider context of indulgence" (Goody 1984, 117). This indulgence was the product of the plow technology generating a surplus that the rich evolved the means to collect.

Similarly, the stratification system of medieval and feudal Europe used food as a class marker. Braudel characterized European history as "two opposed groups: meat eaters (the few) and the rest who eat bread, gruel, roots, and cooked tubers" (1981, 106). While the secular and ecclesiastical upper classes ate meat, "the peasantry's diet was predominantly based on cereals and vegetables, and was often sparse at that" (Mennell 1986, 41).

The European diet was largely tied to wheat. First grown in China (and originally used as vermicelli), wheat spread throughout the world in the nineteenth century. Wheat was well suited to the relatively sparsely populated areas of Europe. By world standards (until the Green Revolution, see chapter 8) "wheat's unpardonable fault is with its low yield: it did not provide for its people adequately" (Braudel 1981, 120). It was not until the 1800s, with the three-field rotation and improved wheat growing, that enough surplus was generated to support such large cities as London.

While the rich ate wheat (which the peasants generally sent to market because it brought a higher price), the peasants ate the more coarse grains, rye and maize. "The peasant ate millet and maize and sold his wheat: he ate salt pork once a week and took his poultry, eggs, kids, calves, and lambs to market" (Braudel 1981, 190).

Combined with the meat diet, Braudel estimates the average caloric intake of the rich in the eighteenth century as between 3,500 and 4,000 (1981, 129). The urban worker averaged 2,000 calories, consuming "bread, more bread, and gruel" (Braudel 1981, 129). The average desirable consumption for today's lifestyle is about 2,500 calories.

While wheat sustained Western civilization, rice was the basis of Asian and Indian civilizations. And "maize . . . sustained the brilliance of the Inca, Maya and Aztec civilizations and semi-civilizations . . ." (Braudel 1981, 158).

The most extreme manifestation of the class differences in food consumption can be found in the medieval banquet. Like some tribal *potlatches* where food is conspicuously wasted and given away, the medieval banquet was characterized by massive quantity. Remarking on the banquets of the court of Burgundy in the fifteenth century, Mennell notes that they involved, "enormous jumbled heaps of food. . . . The meal was composed of several courses, each made up of a large number of items habitually served in a pyramid on a single great dish, among which were often found animals—pigs, calves, even oxen—roasted whole" (Mennell 1985, 39). Even a typical royal dinner, like that for George III (1760–1820) contained "pottager vermicelly, pullets with rice pillaw, fillets of mutton and potatoes, cold chicken and sliced tongue, ham with pease and beans, small turbot and small lobster, quales, artychokes, cherry tart, lambs sweetbreads ragou'd, omlettes en roulade" (Mennell 1986, 125). Vegetables were often despised as food for the lower classes, but meat, because it was very fresh, contained adequate amounts of vitamin C (see chapter 3).

Just as the higher religious castes of India marked their place by vegetarian diets and fasting, so did the Christians of Western Europe. In contrast to the gluttonous, carnivorous diet of the rich, monasteries often insisted on some restriction of meat as well as "temporary abstinence of everybody for weekly and annual fasts" Goody 1984, 139). St. Augustine listed gluttony as one of the seven deadly sins. And "Christian moralists saw in elaborate foods and eating ceremonial a way the devil acquired disciples" (Goody 1984, 139, citing Cosman 1976, 117). The previous Catholic Christian prohibition on eating meat on Fridays is from that tradition.

Summary

Human food consumption began with a "garden of Eden" where small human groups could hunt and gather all the food they needed. Their diets were relatively well balanced, the food was usually plentiful, and the number of hours of labor entailed in getting it was not generally regarded as excessive. From this small egalitarian society came an ethic and a practice of food sharing. Food consumption patterns usually reflect the stratification of a society. Hunters and gatherers were the most egalitarian of any society. And they ate relatively well compared to the lower classes of future societies.

The practice of growing food generated an economic surplus. This initial stratification of society relegated women to the more mundane production roles. Nevertheless, their importance was recognized by increasing cases of matrilineal descent in horticultural societies. The hoe usually made possible only a small surplus and therefore a small degree of social inequality.

Agrarian society is the first of which historians have written records. Writing and other occupational specializations are made possible by the large economic surplus generated in agriculture from the use of the plow. Because it was more productive, the lot of the affluent grew much better. But inequality increased. The poor had a lower status and an accompanying low-nutrient diet. Often regarded as less than human in advanced agrarian societies, the lot of the poor was to be the fodder for the massive construction projects of various agrarian elites. Except when disease made them valuable, the poor were largely expendable. Their usual excess numbers allowed the agrarian elites to treat them as animals and to use up their bodies in labor.

We have already raised some issues that require a knowledge of nutrition. The remainder of this book presupposes even more. Therefore, we devote two chapters to mastering the basics of nutrition in order to lay the groundwork for understanding the number of issues that presuppose this knowledge in the rest of the area of food and society.

REFERENCES

Abrams, H. Leon, Jr. 1979. "The Relevance of the Paleolithic Diet in Determining Contemporary Nutritional Needs." *Journal of Applied Nutrition. 31.* 43–59.

Boulding, Elise. 1976. *The Underside of History: A View of Women Through Time*. Boulder, Colo.: Westview Press.

———. 1977. *Women in the Twentieth Century World*. New York: Saga Publications, Inc.

Braudel, Bernard. 1981. *The Structures of Everyday Life: The Limits of the Possible*. Translated by Sian Reynolds. New York: Harper & Row.

Cosman, E.A. 1976. *Fabulous Feasts: Medieval Cookery and Ceremony*. New York: n.p.

Dahlberg, Frances. 1981. *Women the Gatherer*. New Haven: Yale University Press.

Estioko-Griffin, Agnes, and P. Bion Griffin. 1981. "Woman the Hunter: The Agta." In *Women the Gatherer*, ed. Frances Dahlberg. New Haven: Yale University Press.

Goody, Jack. 1984. *Cooking, Cuisine, and Class*. New York: Cambridge University Press.

Grashuis, Monique. 1989. "Women in Food Production." Classroom report, Aquinas College.

Harris, Marvin. 1985. *Good to Eat: Riddles of Food and Culture*. New York: Simon and Schuster.

Hockett, Charles F., and Robert Ascher. 1964. "The Hunger Revolution." *Current Anthropology 5*: 135–168.

Kraybill, Nancy. 1977. "Pre-Agricultural Tools for the Preparation of Foods in the Old World." In *Origins of Agriculture*, ed. Charles Hockett, 485–521. Chicago: Aldine.

PRO. 1673. *The History of the Coronation of King James II and Queen Mary*. London: n.p.

Lenski, Gerhard, and Jean Lenski. 1982. *Human Societies*. New York: McGraw-Hill.

Mennell, Stephen. 1986. *All Manners of Food*. New York: Basil Blackwell.

Sahlins, Marshall. 1968. *Tribesman*. Englewood Cliffs, N.J.: Prentice Hall.

Sanderson, Stephen K. 1991. *Macrosociology*, 2nd ed. New York: Harper Collins.

Tannahil, Reay. 1984. *Food in History*. New York: Stein and Day.
Zihlman, Adrienne L. 1987. "Women as Shapers of Human Adaption." In *Woman the Gatherer*, ed, Frances Dahlberg, 75–104. New Haven: Yale University Press.

Note

[1] 1.Contemporary meat production is also wasteful of energy. We expend 5–7 calories to create one calorie of meat. Each pound of beef requires 9–15 pounds of grain for its production.

Chapter 3

Scientific Nutrition from the Industrial Revolution

The following food product of the Industrial Revolution lists as its contents:

> corn syrup, enriched flour, niacin (a "B" vitamin), iron (ferrous sulphate), thiamine monitrate (B2), sugar, water, partially hydrogenated vegetable and/or animal shortening (contains one or more of: canola oil, corn oil, cottonseed oil, soybean oil, beef fat), eggs, skim milk, (contains 2% or less of whey), modified food starch, salt, leavening (contains one or more of: baking soda, monocalcium phosphate, sodium acid pyrophosphate), mono- and dyglycerides, leicithin cellulose gum, polysorbate 60, sodium caseinate, natural and artificial flavors, artificial color (yellow 5), sorbic acid (to retard spoilage).

According to Freydburg and Gartner in *The Food Additives Book*, of the above ingredients the following are problematic:

—Artificial color, yellow 5: The ingredient, tartraxzine, "has been found to cause allergic reactions . . . serious on occasion" (1982, 538).

—Modified food starch: "Some diarrhea, some slower growth and some calcium deposition in kidney tubules have occurred with a few of these modified starches" (1982, 646).

—Sodium caseinate: "some of the processes by means of which it [casein] is extracted and converted to caseinate can form lysinoaline (LAL), a substance which has been found to cause kidney damage in experiments with rats" (1982, 5).

—Corn syrup (and sugar): "The principal health concern is dental caries (cavities)" (1982, 520). Because this ingredient is listed first, it is the biggest ingredient by weight.

—Partially hydrogenated vegetable or animal shortening: This will be high in fat, therefore tends to raise LDL (bad) cholesterol levels.

In chapter 1 we noted the decline in the general quality of nutrition (except for a small elite) with the advance of civilization. In the move from hunting and gathering to food-growing societies, the lower classes experienced nutritional deprivation in quantity and quality.

With the Industrial Revolution, food became the object of industrial processes. Whereas previous forms of food preservation were devoted to maintaining food in its natural state—refrigeration, canning, and to some degree salting and smoking—advanced industrial processes involved application of the products of chemistry to food.

The result was often a chemical cornucopia that bore little resemblance to real food. The ingredient description that begins this chapter is filled with artificial color, devitaminized flour sprayed ("enriched") with vitamins, emulsifiers, and preservatives. This product of American industrial creativity is a *Hostess Twinkie*. This "golden sponge cake with creamy filling," which serves as a snack food for millions of Americans, is often displayed at the center of convenience stores. It is the butt of jokes about nutritional worthlessness among scientific dietitians. Is this what industry and science have to offer in the way of food progress?

In part, this question raises again the issues of oligopolization in a capitalist socioeconomic system. But before tracing the influence of profit-oriented patterns of ownership (chapter 4), I review the development of academic scientific nutrition.

Until this century, the scientific knowledge of food was sketchy at best. Vitamin C was discovered after sailors brought limes or lemons onto English ships. The sailors did not get the scurvy that others did. The name "limey" comes from this discovery. Similarly, B vitamins were discovered because some chickens that were temporarily being fed polished white rice got beriberi. When chickens were fed brown rice, Christian Eijkmans noticed that they recovered. Drs. Eijkmans and Grinjins discovered the element (B vitamin) in the bran that caused the recovery.

As we now know, food is composed of five basic elements: protein, fat, carbohydrates, vitamins, and minerals. I will discuss each briefly.

Protein

Known agrarian civilizations have, mostly by chance and taste, hit on combinations of vegetable goods that compose a complete protein. Recent research indicates that complementary proteins need not be eaten at the same meal. Variety during the day is sufficient.

Protein is composed of twenty-two different amino acids. The body makes thirteen of them, but nine must be obtained from food. The trick to a successful vegetarian diet (typical of the masses of most agrarian societies) is to combine vegetable proteins that "complement" each other. Bryant et al. cite beans and corn as a typical example. "Beans are low in the amino acid methionine and high in lysine. Corn has the opposite amino acid pattern" (1985, 45). When beans and corn are combined (along with their other seven amino acids), as they are in a Mexican corn tortilla with beans in it, one gets complete protein.

Some of the other traditional combinations which lead to complete protein are:

Legumes (soybeans, peanuts, black-eyed peas, kidney beans, chickpeas, navy beans, pinto beans, lentils, split peas, lima beans	with	Corn, rice, wheat, sesame seeds, barley, oats. (Brody 1987, 37)

In addition, any incomplete protein is completed by the addition of a complete protein. The addition of a small amount of meat (fowl or fish) completes any incomplete protein. Dairy does the same.

The reader will recognize here the basic pattern of two of the most sophisticated cuisines in the world, the Chinese and Indian. Both mix a small amount of flesh with the basically vegetarian. As T. Colin Campbell of Cornell University reports from the China Project, "a diet made up of at least 80% to 90% plant materials may be optimal" (Liebman 1990a, 5).

In 1910, Americans ate a reasonably well balanced diet. Since then, the Center for Science in the Public Interest reports "a 34% increase in fat consumption . . . a 100% increase in consumption of sugar and other caloric sweeteners . . . a 40% decrease in complex carbohydrates" (Brody 1987, 13).

Fat

In addition to protein, food is composed of fats and carbohydrates. Of fat, we need about a tablespoon per day. Additional fats add to obesity, clogged arteries and heart failure. (Yet today's fast-food method of frying adds to fat.) Fat contributes to the cholesterol level in the blood. Cholesterol builds up in fatty deposits on the walls of the arteries, causing decreased passage to the heart. This contributes to heart attacks and strokes.

There are three varieties of fats: saturated, monounsaturated, and polyunsaturated. Although all contribute calories and body fat, the saturated fats raise the level of blood cholesterol, while mono and polyunsaturated fats generally lower it. Saturated fat comes mostly from animal sources, but also from palm and coconut oil. Included are butter, cheese, chocolate, egg yolk, lard, meat, poultry (mostly the skin), and the kind of vegetable shortening that is solid at room temperature.

The better (but no less fattening) kinds of fats are mono and polyunsaturated. Monounsaturated products (not quite the best) are avocado, cashews, olives and olive oil, peanuts and peanut oil, and peanut butter. The best, polyunsaturated, are almonds, corn oil, cottonseed oil, filberts, fish, most margarine, pecans, safflower oil, soybean oil, sunflower oil, walnuts, and canola oil (Brody 1987, 63).

There is increasing evidence that in countries where people eat diets high in dietary fat there is a large amount of colon cancer. Reviewing a number of studies, the Center for Science in the Public Interest notes a substantial link between the incidence of colon cancer and the eating of high fat diets throughout the world. In the past thirty years, those Japanese who have converted to Western high fat and meat diets have evidenced a colon cancer death rate increase of 130 percent (Liebman 1990, 5).

Meat also poses a risk of chemical contamination. Before the petrochemical aspect of the Industrial Revolution, almost all farming was what we now term *organic*. Manure and humus (the product of composting) were the basic fertilizers. Scarecrows and flapping cloth kept some pests away. Today, America has the most chemical-intensive farming practices in the world. Largely due to a profit orientation (see chapter 8), contemporary agriculture uses petrochemical sources of fertilizer, pesticides, herbicides, and fungicides. As David Pimental of Cornell University reports, "in the last 40 years, the volume of pesticides used in American agriculture has increased tenfold" (Montgomery 1987, 7). In addition, antibiotics and growth hormones are added to the food of animals.

If chemicals are used on crops, some of them inevitably remain with the food. People that eat the food consume some of the pesticides. Usually they are in "tolerable" amounts. But a cow must consume nine to sixteen pounds of grain to create a pound of beef. With that grain poundage come the chemicals. To eat beef is to eat "high on the food chain." Rather than the chemicals in one pound of grain, a pound of beef can contain most of the chemicals in nine to sixteen pounds of grain. The result is a degree of chemical contamination that can pose serious health risks to the consumer. Leslie Goodman-Malamuth cites the results of a congressional subcommittee that said, "nearly 90% of the 20,000 to 30,000 animal drugs currently in use are not FDA approved."[1].

Of particular concern is the increasing role of the fast food industry in feeding Americans. There has been some improvement partly due to negative advertising.[2] But, increasingly, the fast-food regimen of a diet of high fat, sugar, and low carbohydrates has influenced the eating patterns of those who would simulate the heyday of American upper-class steak eating in the 1950s. For many of America's poor, a trip to McDonald's is the closest they get to "eating out."

At the same time, the increased demand for beef is a major factor in the decline of rain forests in much of the developing world. Because of the demand for more farmland to feed the cows, large landowners are cutting down potentially invaluable biological resources to allow further grazing land for cattle. While this meat is usually too expensive for the local population (which often works on the farms and in the slaughterhouses), it is shipped to the highest bidder in the "developed" northern world (see chapter 4).

What has been the role of government in this nutritional snarl? In seeking answers to this question one must raise issues about the role of government generally in a capitalist industrial society.

According to James O'Connor (1973), the two primary functions of government in a capitalist industrial society are *accumulation* and *legitimation*. That is to say, rather than act as a neutral umpire (as some high school civics texts would have us believe), government actively seeks to help profit-making enterprises thrive. It does this by often providing the infrastructure on which private business can profit. In the automobile industry, this was the highway system. In food, it often takes the form of agricultural subsidies to the Agricultural Experiment and Extension Services. The state sponsors agricultural agents who promote a capital-intensive (chemical) approach to farming, sponsorship of research in biotechnology, lax regulations in areas where strict regulation would cost private industry too dearly, and policies toward labor that generally favor the owners.

Nevertheless, in the kind of democracy that exists in the United States, government must at least give the appearance of acting in the interests of consumers. Therefore, in the area of food, it regulates the use of toxins in food production and consumption. In actual fact, it regulates badly. Only by confronting massive evidence did it ban the pesticide DDT. Yet it still allows the manufacture of that pesticide and its sale to many developing nations that sell produce on which DDT has been used to us (Weir and Shapiro 1981). This pattern holds for a number of other toxic and semitoxic chemicals.

For instance, the Food and Drug Administration inspects only about 4 percent of the imported food. One can hardly think of our food supply as safe. "It doesn't test for about half the pesticides that are present. And it doesn't sample randomly" (Lefferts 1989, 5). In addition, the Food Safety and Inspection Service does not test meat and poultry for "80% of the pesticides that could be present" (Lefferts 1989, 6).

But because it is easier to recommend than to regulate, the U.S. government has made recommendations about what Americans should be eating. Twenty years ago, this was confined to the "four food groups": meat (poultry, fish, meat and eggs), 2 servings per day; dairy, 3 to 4 servings per day; beans, grains and nuts, 4 or more servings per day; and fruits and vegetables, 4 or more servings per day. This division is the one that has been (and may still be) taught by schoolteachers.

Many dietitians continue to approve the eating of red meat in spite of all the evidence that red meat is deleterious to health. Public school systems continue to accept contributed literature from meat and dairy producers that emphasizes this traditional pattern (Belasco 1989, 129).

It was therefore a bit progressive when, in 1977, Senator George McGovern's Senate Select Committee on Nutrition and Human Needs made some recommendations that moved away from the "accumulation" function of government and toward "legitimation" (the interest of the general people) (Belasco 1989, 150–153). What follows are the McGovern recommendations now generally endorsed by the mainstream of American dietitians (in word, if not always in deed):

1. Increase carbohydrate consumption to account for 55% to 60% of the energy (caloric) intake.
2. Reduce overall fat consumption from approximately 40% to 30% of energy intake.
3. Reduce saturated fat consumption to account for about 10% of total energy intake and balance that with polyunsaturated and monounsaturated fats, which account for about 10% of energy intake each.
4. Reduce cholesterol consumption to about 300 mg. a day.
5. Reduce sugar consumption by about 40% to account for about 15% of total energy intake.
6. Reduce salt consumption by about 50% to 85% to approximately 3 grams a day.

This led to the recommendation for 2 meat-group servings (total 6 oz.); 6–11 servings from beans, nuts and grains; 2–3 servings from dairy; 3–5 servings from vegetables; 2–4 servings from fruits ("Dietary Guidelines" 1991, 4).

Carbohydrates

The government recommendation is to increase carbohydrate consumption. Carbohydrates are plant foods. The unrefined kind (white sugar is highly refined) are whole grains, beans, fruits, and vegetables, "the only food category not linked to any of the leading killer diseases" (Brody 1987, 94–95).

Copyright June 1991, Center for Science in the Public Interest. Reprinted from *Nutrition Action Health Letter* (1875 Connecticut Avenue NW, Suite 300, Washington, DC. 20009-5728. $20.00 for 10 issues), p.3.

These are the basic foods that, after the demise of hunting and gathering societies, human beings learned to grow. In addition to the domestication of animals for meat and milk, complex carbohydrates have formed the basis of the human diet since 3000 B.C. And, previous to the Industrial Revolution, their production was all by "organic" techniques and animate (human and animal) labor.

In their refined form, carbohydrates are fattening and bereft of vital nutrients. We encounter refined sugars and starches in a multitude of products that prey on our predilection for sweets. The Twinkie is one of the worst examples. Nutritionists refer to this kind of carbohydrate as "empty calories." In addition to all its chemicals, it includes refined sugar and the white, refined flour from which the bran and B vitamins have been removed.

Complex carbohydrates (grains, legumes, vegetables, and fruits) are generally low in calories and high in nutrient values. We have already examined how they may be combined to form complete vegetable proteins. Furthermore, complex carbohydrates generally contain a high ratio of fiber. Lacking most nutrients, fiber is usually indigestible. Its role is to push through the digestive tract, cleaning it so that such remains as meat do not putrefy and lead to cancer. It reduces constipation so that the "internal plumbing" functions smoothly.

Some dietary fiber has been touted as chemically active in the fight against cholesterol as well. For the past few years oat bran has held this distinction. It has even been added to some of the Hostess products. Other candidates for this honor include rice bran. But the nutritional profession is not entirely certain whether soluble fiber is the reason for lower rates of cancer or whether fiber's role in flushing out the digestive system is its main contribution.

If the entire population of the developed world ate more complex carbohydrates and less meat, less grain would be used to produce meat. That grain could be eaten. Were this to happen, the amount of usable food would be significantly increased, probably enough at the world level to feed the hungry everywhere (see chapter 9).

Vitamins and Minerals

In addition to protein, fat and carbohydrates, the human body needs specific micro nutrients. Recall the experiments with scurvy among English "limey" sailors and the way that beriberi was solved in the chicken experiment. Over the past few decades, dietitians in controlled laboratory settings have discovered a great deal about vitamins. Yet much popular ignorance and confusion remains.

At present, the known vitamins are generally divided into water-soluble and fat-soluble vitamins. The water-soluble ones—generally the Bs and C—are needed daily. Fat-soluble vitamins—A, D, E, and K—cannot be flushed easily. They can be stored in the body and can accumulate to excess. One can, in extreme cases, get too much of any fat-soluble vitamin. Toxicity is the result. With vitamin A (beta carotene), there is now serious disagreement about the upper limits of tolerance (Liebman 1988s, 6). Nevertheless, megadoses of vitamins A, D, E, and K can be toxic. Table 3.1 lists the vitamins and their sources.

In addition to vitamins, there are a variety of minerals that the human body must have to function optimally. These include calcium, phosphorous, magnesium, potassium, sulfur, chloride, iron, copper, zinc, iodine, fluorine, chromium, selenium, manganese, and molybdenum.

If one eats routinely from the four food groups (with less from the meat group and more from complex carbohydrates), one should obtain virtually all the vitamins and minerals which are necessary to sustain optimal human life. But many people slight good food and eat unbalanced meals. In addition, bodies vary enormously. The large number of reported success with megavitamin doses that characterize the popular *Prevention Magazine* gives evidence of the idiosyncratic nature of human ills and cures. Although many of these cases are not supportable by controlled experiments (including a lot of the claims Linus Pauling makes for vitamin C megadoses), there are legitimate successes in some cases. One must work with one's body while staying beneath the tolerable limits of various vitamins and minerals.

Inevitably the question arises about the "placebo effect" on many individuals who claim success with vitamins and minerals. Does the fact that one believes what one is doing will make her or him better actually influence the process of healing? Numerous studies document the effectiveness of placebos. An estimated one-third to one-half of people in many situations get better from their administration. Is it not possible that this may be true with many of the claims about vitamin and mineral cures?

At the moment there is no definitive answer to this question. If the placebo effect is one way in which the mind influences healing, however, would a minimal attention to vitamins and minerals be a conscious attempt to use the mind's power to heal? If the expense is not enormous, one might engage in a daily ritual of up to 3000 mg of vitamin C per day, or take a multivitamin. If one does not substitute vitamins for correct eating, is it not useful to engage the mind's power to heal or to prevent illness? Is that not what a lot of traditional healing was about? And if the supplement is purchased cheaply and actually works for one's body, so much the better.[3]

The position one should avoid is substituting one's own "therapy" for that of the medical establishment. While it is certainly true that the medical profession has many flaws, it still provides the best general body of knowledge in the world about the human body. One may supplement its knowledge with the traditional therapies of the world (including vitamins and minerals), but one runs serious risks by substituting a home-administered regimen of megadoses of micronutrients, especially in areas where standard Western medicine has traditionally been successful.[4]

Summary

This chapter has considered the results of the scientific knowledge of nutrition in the twentieth century. Health and medical science have discovered the reasons for the usefulness of

Table 3.1 Vitamins

	Major Food Sources	Adult RDA*	What It Does	Potential Benefits	Supplementation
A	Milk, eggs, liver, cheese, fish oil. Plus fruits & vegetables that contain beta carotene. (You need not consume preformed vitamin A if you eat foods rich in beta carotene)	800 RE (8,000 IU), women; 1,000 RE (10,000 IU), men 1 C. milk = 140 RE	Promote good vision; helps form & maintain skin, teeth, bones, & mucous membranes. Deficiency can increase susceptibility to infectious disease.	May inhibit the development of certain tumors; may increase resistance to infection in children.	Not recommended, since toxic in high doses.
Beta carotene (not a vitamin, but converted to vitamin A in the body)	Carrots, sweet potatoes, cantaloupe, leafy greens, tomatoes, apricots, winter squash, red bell peppers, pink grapefruit, broccoli, mangos, peaches.	No RDA; experts recommend 5–6 mg (milligrams) 1 med. carrot = 12 mg 1 sweet potato = 15 mg	Converted into vitamin A in the intestinal wall. As an antioxidant, it combats the adverse effects of free radicals in the body. Best known of a family of substances called carotenoids.	May reduce the risk of certain cancers as well as coronary artery disease.	6–15 mg (equal to 10,000–25,000 IU of vitamin A) a day for anyone not consuming several carotene-rich fruits or vegetables daily. Nontoxic.
C (ascorbic acid)	Citrus fruits & juices, strawberries, tomatoes, peppers (especially red), broccoli, potatoes, kale, cauliflower, cantaloupe, Brussels sprouts.	60 mg 1 orange = 70 mg 1 C. fresh O.J. = 120 mg 1 C. broccoli = 115 mg	Helps promote healthy gums & teeth; aids in iron absorption; maintains normal connective tissue; helps in healing of wounds. As an antioxidant, it combats the adverse effects of free radicals.	May reduce the risk of certain cancers, as well as coronary artery disease; may prevent or delay cataracts.	250–500 mg a day for anyone not consuming several fruits or vegetables rich in C daily & for smokers. Larger doses may cause diarrhea.
D	Milk, fish oil, fortified margarine; also produced by the body in response to sunlight.	5 mcg (micrograms), or 200 IU; 10 mcg or 400 IU, before age 25 1 C. milk = 100 IU	Promotes strong bones & teeth by aiding the absorption of calcium. Helps maintain blood levels of calcium & phosphorus.	May reduce the risk of osteoporosis.	400 IU for people who do not drink milk or get sun exposure, especially strict vegetarians & elderly. Toxic in high doses.

Table 3.1 Vitamins (Continued)

	Major Food Sources	Adult RDA*	What It Does	Potential Benefits	Supplementation
E	Vegetable oil, nuts, margarine, wheat germ, leafy greens, seeds, almonds, olive, asparagus.	8 mg women; 10 mg men (12–15 IU) 1 T. canola oil = 9 mg 1 T. margarine = 2 mg 1 oz. peanuts = 2 mg 1 C. kale = 6 mg	Helps in the formation of red blood cells & the utilization of vitamin K. As an antioxidant, it combats the adverse effects of free radicals.	May reduce the risk of certain cancers, as well as coronary artery disease; may prevent or delay cataracts; may improve immune function in the elderly.	200 to 800 IU advised for everybody; you can't get that much from food, especially on a low-fat diet. No serious side effects at that level, though diarrhea & headaches have been reported.
K	Intestianl bacteria produce most of the K needed by the body. The rest is supplied by leafy greens, cauliflower, broccoli, cabbage, milk, soybeans, eggs.	60–65 mcg women; 70–80 mcg men 1 C. broccoli = 175 mcg 1 C. milk = 10 mcg	Essential for normal blood clotting.	May help maintain strong bones in the elderly.	Not necessary, not recommended.
Thiamin (B1)	Whole grains, enriched grain products, beans, meats, liver, wheat germ, nuts, fish, brewer's yeast.	1–1.1 mg women; 1.2–1.5 mg men 1 pkt oatmeal = 0.5 mg	Helps cells convert carbohydrates into energy. Necessary for healthy brain, nerve cells, & heart function.	Unknown	Not necessary, not recommended.
Riboflavin (B2)	Dairy products, liver, meat, chicken, fish, enriched grain products, leafy greens, beans, nuts, eggs, almonds.	1.2–1.3 mg women; 1.4–1.7 mg men 1 C. milk = 0.4 mg 3 oz. chichen = 0.2 mg	Helps cells convert carbohydrates into energy. Essential for growth, production of red blood cells, & health of skin & eyes.	Unknown	Not necessary, not recommended.

Table 3.1 Vitamins (Continued)

	Major Food Sources	Adult RDA*	What It Does	Potential Benefits	Supplementation
Niacin (B3)	Nuts, meat, fish, chicken, liver, enriched grain products, dairy products, peanut butter, brewer's yeast.	13–19 mg 3 oz. chicken = 12 mg 1 slice enriched bread = 1 mg	Aids in release of energy from foods. Helps maintain healthy skin, nerves, & digestive system.	Large doses lower elevated blood cholesterol	Megadoses may be prescribed by doctor to lower blood cholesterol. May cause flushing, liver damage, & irregular heart beat.
B6 (pyroxidine)	Whole grains, bananas, meat, beans, nuts, wheat germ, brewer's yeast, chicken, fish, liver.	1.6 mg women; 2 mg men 1 banana = 0.7 mg 1 C. lima beans = 0.3 mg	Vital in chemical reactions of proteins & amino acids. Helps maintain brain function & form red blood cells.	May boost immunity in the elderly.	Megadoses can cause numbness & other neurological disorders.
B12	Liver, beef, pork, poultry, eggs, milk, cheese, yogurt, shellfish, fortified cereals, fortified soy products.	2 mcg 1 C. milk = 0.9 mcg 3 oz. beef = 2 mcg	Necessary for development of red blood cells. Maintains normal functioning of nervous system.	Unknown	Strict vegetarians may need supplements. Despite claims, no benefits from megadoses.
Folacin (a B vitabin; also called folate or folic acid)	Leafy greens, wheat germ, liver, beans, whole grains, broccoli, asparagus, citrus fruit & juices.	180 mcg women; 200 mcg men 1 C. raw spinach = 110 mcg 1 pkt oatmeal = 150 mcg 1 C. asparagus = 180 mcg	Important in the synthesis of DNA, in normal growth, & in protein metabolism. Adequate intake reduces the risk of certain birth defects, notably spina bifida.	May reduce the risk of cervical cancer.	400 mcg from food or pills, for all women who may become pregnant, in order to help revent birth defects.
Biotin (a B vitamin)	Eggs, milk, liver, brewer's yeast, mushrooms, bananas, tomatoes, whole grains.	No RDA; experts recommend 30–100 mcg	Important in metabolism of protein, carbohydrates, & fats.	Unknown	Not necessary, not recommended.
Pantothenic acid (B5)	Whole grains, beans, milk, eggs, liver.	No RDA; experts recommend 4–7 mg	Vital for metabolism of food & production of essential body chemicals.	Unknown	Not necessary, not recommended. May cause diarrhea.

* These figures are not applicable to pregnant women, who need additional vitamins and should seek professional advice.

Source: Excerpted from the University of California at Berkley Wellness Letter, Health Letter Associates, January 1994.

many traditional diets. Cultures such as those of India and China have evolved diets that, by combining complementary proteins, have successfully nourished their populations. Societies that did not develop such successful diets are no longer available, for the most part, for study.

As we now know it, food is composed of protein, carbohydrates, fats, vitamins, and minerals. One should be able to obtain these nutrients from eating from the four food groups regularly. Nevertheless, in the light of our most modern medical research, one should be aware that our food system is run mostly by oligopolies for short-term profit. As a result, they have developed some technologies that are potentially destructive to our health. That subject and the manner in which one can defend against them is discussed in chapter 4.

References

Belasco, Warren. 1989. *Appetite for Change.* New York: Pantheon.

Brody, Jane. 1987. *Jane Brody's Nutrition Book.* New York: Bantam.

Bryant, Carol, Anita Courtney, Barbara Markesbery, and Kathleen Dewalt. 1985. *The Cultural Feast: An Introduction to Food and Society.* New York: West.

"Complete Protein: An Incomplete Theory." 1994. *Consumer Reports on Health,* January, 2–3.

"The Dietary Guidelines Become More User Friendly." 1991. *Tufts University Diet and Nutrition Letter,* January, 5.

"Eschewing the Fat." 1991. *Harvard Health Letter,* March, 1–4.

Fieldhouse, Paul. 1986. *Food and Nutrition: Customs and Culture.* New York: Croom Helm.

Freydberg, Nicolas, and Willis A. Gartner. 1982. *The Food Additives Book.* Mt. Vernon, N.Y.: Consumers Union.

Friedland, William. 1989. "Is Rural Sociology Worth Saving?" *Rural Sociologist,* Winter, 3–6.

Goodman-Malamuth, Leslie. 1985. "Junk Food Goes Hollywood." *Nutrition Action Health Letter,* September/October, 1, 4–7.

———. 1986a. "Fast Food and Kids." *Nutrition Action Health Letter,* March, 1, 4–7.

———. 1986b. "Animal Drugs." *Nutrition Action Health Letter,* May, 1, 5–7.

Harris, Marvin. 1985. *Good to Eat*. New York: Simon and Schuster.
Hegstead, D. Mark. 1978. "U.S. Dietary Goals." Talk at Food and Agriculture Outlook Conference.
League of Women Voters. 1989. *America's Growing Dilemma: Pesticides in Food and Water*. Washington, D.C.: League of Women Voters Education Fund.
Lefferts, Lisa. 1989. "Pass the Pesticides." *Nutrition Action Health Letter*, April, 1, 5–7.
Liebman, Bonnie. 1988a. "Carrots against Cancer." *Nutrition Action Health Letter*, December, 1, 5–7.
———. 1988b. "Fatty Complexes." *Nutrition Action Health Letter*, September, 1, 5–7.
———. 1990a. "The Changing American Diet." *Nutrition Action Health Letter*, May, 8–9.
———. 1990b. "Lessons from China." *Nutrition Action Health Letter*, December, 1, 5–7.
McLeod, J.C., and A.A. Jackson. 1982. "Rehabilitation of a Malnourished Rastafarian Child." *Cajuns ANP/HNF 14*(4): 202–209.
Mauer, Donna. 1989. "Becoming a Vegetarian: Learning a Food Practice and Philosophy." Paper presented at the meeting of the Association for the Study of Food and Society, College Station, Texas, June.
Mintz, Sydney. 1985. *Sweetness and Power*. New York: Viking.
Montgomery, Anne. 1986. "You are What You Eat: Anthropological Factors Influencing Food Choice." In C. Ritson and L. Fofton, *Chiththento*, Sussex, UK: Wiley.
O'Connor, James. 1973. *The Fiscal Crisis of the State*. New York: St. Martin's Press.
"Preventing Cancer." 1987. *Nutrition Action Health Letter*, July/August, 1, 4.
Robbins, John. 1987. *Diet for a New America*. Walpole, N.H.: Stillpoint.
"Simply Sugar or More Sugar." 1988. *Tufts University Diet and Nutrition Letter*, July, 1.
"U.S. Dietary Goals." 1977. *Nutrition Review*, May, 122–125.
Visser, Margaret. 1986. *Much Depends on Dinner*. New York: Grove Press.
Weir, David, and Mark Shapir. 1981. *Circle of Poison*. San Francisco: Institute for Food and Development Policy.
Zucherman, Sam. 1985. "Antibiotics: Squandering a Medical Miracle." *Nutrition Action*, January/February, 8–11.

Notes

[1] 1.Further, "the 'potentially significant adverse affects' of as many as 4,000 of these products may be passed on to humans via drug residues in the milk, meat and eggs we consume" (Goodman-Malamuth 1986, 5).

[2] 2.In the spring of 1990, industrialist and consumer health advocate Phil Sokolof of Omaha, Nebraska, funded full-page advertisements in twenty newspapers, including the *Wall Street Journal* and *USA Today*, attacking the saturated fats in McDonalds beef. "The ad said . . . McDonalds' Big Mac and a bag of french fried contained 25 grams of saturated fats." *Grand Rapids Press*, April 5, 1990, A6.

[3] 3.Vitamin and mineral supplements are mostly unregulated and generally very profitable for the companies that make them. They are usually obtained most cheaply from national mail-order companies.

[4] 4.Many "traditional" therapies compete with Western medicines and are generally less profit oriented. These include Chinese and Native American herbalism, acupuncture, Hispanic and Asian "hot" and "cold" food therapies, and a number of therapies that involve using hands to heal.

Chapter 4

Food, Profit, Nutrition—A Remedy

> . . . . in the seventies, the food companies got scared. For one thing, there was the space crunch. Many executives wondered if the human stomach could hold more than the 1,400 pounds of food products people ate annually; this total had increased only 5% in the 1960's and held virtually stable through the 1970's and 1980's. "People will buy only so much food," complained one Nabisco executive. To top it off, consumerist pressure was forcing the FDA to take a harder look at some of the chemicals going into new food products. For the already cautious managers of food companies it seemed logical to seek growth not through costly internal R & D but through other, less risky channels: acquisitions of the most profitable small companies, overseas trade school lunch programs, making essentially minor modifications in existing lines, and stretching out the "life cycle" of successful items.
>
> —Warren Belasco, 1989

In the introduction and chapter 1, I raised the issue of how the form of ownership and technology influence food. In a profit system, food producers and marketers must generate a profit. In the early (entrepreneurial) stages of a capitalist food system, the means of competition were increasing the quality and/or decreasing the price. This generally worked to the advantage of the consumer.

But, when the dominant dynamics are those of oligopoly, capitalist competition is through advertising and product proliferation. Price, influenced predominantly through "price leading," almost always goes up, and innovations in quality are often cosmetic rather than substantial. In addition, capitalism seems an inherently unstable system (Baran and Sweezy 1966). Firms must be either expanding or dying.

In our oligopolistic situation, food companies are faced with the dilemma of expanding food purchasing in the context of approximate limits of food consumption. Whereas one can buy innumerable cars, the size of the human stomach limits food consumption.

Similarly, if the major concern is short-term profit (in contrast to long-term corporate planning in the German and Japanese markets), then one must seek the least expensive means to generate profit while maintaining the illusion that the new foods are "good" for the consumer. This has led to a situation where, "of seventeen major industries, food ranked fourteenth in R & D, and of this . . . 80%–90% was devoted to product differentiation rather than research on substantially new food" (Belasco 1989, 189).

Although Western science has taught us most of the basics of good nutrition (see chapter 3), the form of ownership generally works against nutritional health. Large profits are generally not to be made in healthy plain fruits, vegetables, and grains (complex carbohydrates). Rather, high profits are made in highly processed products, heavily advertised and often of little nutritional value. The Hostess Twinkie is but one such example.

This chapter is concerned with the manner in which profit motives outweigh those of basic nutrition. Sugar and meat provide us with examples of product manipulation. And a look at government attempts to regulate harmful chemical additives demonstrates the hegemony of the accumulation function of the state—often at the expense of its legitimation (populist/health) function.

Finally, this chapter examines strategies that food consumers have adopted to protect themselves from the deleterious effects of the food generally produced in our economic system. In this regard, I examine vegetarianism as a self-defense strategy—among its other justifications.

Sugar

Contained within the story of sugar is a great deal of insight into the manner in which capitalist industrialism developed. Sugar figures dominantly in the production, distribution, and consumption aspects of our form of civilization.

The earliest history of sugar is of minimal import. The Greeks had no word for it. The Romans called it "saccharum" (from which comes our modern "saccharin") and used it medicinally—like honey. Sugar played a moderate role in the Islamic world. Dufty raises questions about its possible place in the decline of that world (1975, 30). Sugar came to the West through the Crusades. Because of its initial expense, its use was confined to the rich. It was used as a spice, a medicine, and a preservative.

As sugar production declined in the areas previously controlled by Arabs (the eastern Mediterranean), the prospects for profit grew in the western Mediterranean. When slave labor became available through the triangular trade with the Caribbean islands, investors made sugar a vehicle of capital accumulation. Slave labor in the Caribbean soon killed the incipient sugar industry in Spain and Portugal.

Less well known than the famous triangular trade of cotton, cloth (and manufactured products), and slaves for the mainland was this second triangular trade. English investors sent ships filled with rum to the African coast, exchanged the rum for slaves, sent the slaves to the Caribbean, and took on molasses to be shipped back to England and processed into rum. Each step in the process engendered profit, and if the trip were successful, it could make these entrepreneurs rich.

Sugar consumption came to be intimately tied to the English character. Originating with the rich, the combination of addictiveness of the habit, the desirability of imitating the rich, the profit to be made in importing sugar, and the addition of sugar to a wide variety of English dishes (tea and preserves are the most important) made sugar's rise to a dominant position in English life seem almost inevitable (Mintz 1985, 39). The English government (and navy) established the British Empire to the benefit of entrepreneurial capitalists (and the glory of England). The reader should recognize the role of the state (government) as accumulation—a kind of socialism for the rich while preserving the competitive market in labor for the English working poor. In the Caribbean, slave labor on sugar plantations was the method of accumulation. Colonial governments were guarantors of order.

While sugar had previously been available only to the rich, a "taste" for it arose when the price made it possible for the poorest classes to afford it. By 1750 Mintz reports that it was generally available for use in tea.[1] Its function as the mark of the rich declined, while its role as a source of profit increased (Mintz 1985, 95).

Sugar played a more minor role as a medicine, dentifrice, and decoration (as cake icing). By the nineteenth century, controversies undermined certainty about its medicinal role. But it was used as sweetener and a preservative on a mass scale.

> Sugar as a sweetener came to the fore in connection with three other exotic imports—tea, coffee, and chocolate—of which one, tea, became and has since remained the most important non-alcoholic beverage in the United Kingdom . . . All contain stimulants and can be properly classified as drugs.

> What is interesting about tea, coffee, and chocolate—all harshly bitter substances that became widely known in Great Britain at approximately the same time—is that none has been used exclusively with a sweetener in its primary cultural setting. (Mintz 1985, 109).

Tea in China, coffee in North Africa and the Middle East, and chocolate in its American homeland were all drunk plain and, except for chocolate, often still are. In fact the Mesoamerican ruler Montezuma was reputed to have drunk fifty goblets of unsweetened chocolate before entering his harem (Knapp, 1920, 7). Chocolate's use as an aphrodisiac predated its combination with sugar.

During the seventeenth and eighteenth centuries historians noted the decline of nutrition among the working classes coincident with the increased profits from the selling of tea and sugar to them; "the substitution of tea for beer was a definite nutritional loss; tea was bad not only because it was a stimulant and contained tannin, but also because it supplanted other, more nutritious foods" (Mintz 1985, 117). Because beer was then made without refined sugar, it had been a significant nutritional part of the British diet.

From 1700 to 1750 the nutritional status of the working class declined while its purchasing power increased. Sweet items—the now traditional British puddings, cakes, pastries—made up the difference. Unfortunately, in imitating the eating habits of the rich (who could afford meat) the working class was undermining its own nutritional status. Sugar also entered the working class diet via jam (preserves). Because of sugar's preservative qualities, fruit could be preserved (and sweetened) in jam. This guaranteed a source of vitamin C to the working class.

With the increase in working mothers in the British factory system, sugar once again presented itself as a solution. Mintz writes:

> There seems no doubt that sugar and its by-products were provided unusual access to working-class tastes by the factory system, with its emphasis on saving time, and the poorly paid but exhausting jobs it offered women and children. The decline of bread baking at home was representative of the shift from a traditional cooking system, costly in fuels and in time, toward what we would now proclaim as "convenience eating." Sweetened preserves, which could be left standing indefinitely without spoiling and without refrigeration, which were cheap and appealing to children, and which tasted better than more costly butter with store purchased bread, outstripped or replaced porridge, much as tea had replaced milk and home brewed beer... .
>
> Hot tea often replaced hot meals for children off the job, as well as for adults on the job (1985, 130).

In the past century and a half in the United States, sugar consumption per capita has tripled from around 38 pounds in the 1830s to nearly 100 pounds per capita. In addition, other sweeteners (corn syrup and synthetic sugar substitutes) have come to contribute additional empty calories to our diet.

If sugar contributed to the success of capitalism and the decline of the working-class diet in the process of "development," how has its role changed in our contemporary oligopolistic world? The first element to consider is the existence of a world market and an international sugar trade. Even as the East India Company was the first multinational corporation, so contemporary multinational corporations traverse the globe in search of the cheapest prices on the world market. Today, vertically integrated corporations may control the growing of sugar from creation of the seeds to marketing of the sugar.

More significantly, sugar has been part of the ingredient package (along with many products of industrial chemistry), which adds "convenience" to eating. Just as the industrial process forced the working class into a need for faster food (by taking both parents out of the homes), we now find corporate planners destroying the traditional middle-class family meal with nutritional garbage—usually laden with a high amount of sweetener.

> The food technologist interested in selling products aims willy-nilly at the obliteration of . . . schedules . . . making it possible for everyone to eat exactly what he or she wants to eat, in exactly the quantities and under exactly the circumstances (time, place, occasion) he or she prefers. Incidental to this is the elimination of the social significance of eating together. (Mintz 1985, 201).

Because sugar is partially addictive, the fast-food restaurant industry—dominated by McDonald's—has oozed sugar into as many products as possible. A trip to McDonald's reveals added sugar and other sweeteners on the french fries and hash brown potatoes, in the salad dressing, in the soft drinks and shakes, in the bun for the hamburger, in cookies and pies, and in the Filet-O-Fish, pork sausage, Canadian style bacon, English muffins, danish pastry, and sundae toppings. In addition, it is in the condiments: barbecue sauce, Big Mac sauce, hot cake syrup, hot mustard sauce, ketchup, mayonnaise, strawberry preserves, sweet and sour sauce, tarter sauce (Jacobson and Fritschner 1986, 188–195). As industrialized eating replaces the family meal and much of eating at home (where an adult can serve as "gatekeeper" to family food choice), fast-food restaurants and convenience stores thrive on sugar addiction and sugar profits.

Ultimately the dilemma posed at the beginning of this chapter—how to make a greater profit from a limited amount of possible consumption—has led to a decline in the nutritional status of the industrialized American population. Contemporary obesity often masks actual malnourishment (lack of basic nutrients). Unlike ancient obesity—the result of overindulgence in nutrient-laden foods—contemporary obesity is often the result of high sugar consumption together with a sedentary lifestyle. It seems that oligopolies profit from nonnutrition while different medical oligopolies profit from the extraordinarily expensive cures developed to heal the problems created by the form of nutricidal consumption we have adopted.

The problem has been compounded by lax government regulations about listing ingredients. The new federal requirements may help.

For a long time there has been some controversy about marketing to children. Consumer groups have focused on Saturday morning television programs. In a recent study of them, dietitian Nancy Cotugna of the University of Delaware found that in the twelve hours of commercials she taped, "more than 70% of the 225 commercials were for food products of low nutritional value (pre-sweetened cereals and fruit drinks, high sodium canned pastas, or fatty fast food burger, fries and shakes)!" (in Schmidt 1989, 7).

In classic oligopolistic fashion, food companies seek to compete, not with new and better products of which the consumer can easily judge the healthfulness, but through product proliferation and advertising. The most insidious recent maneuver is to utilize the noncommercial medium—movies—to place products in prominent positions. For example, of Reese's Pieces in the film *E.T.*, Hershey company spokesman Brian Herman said, "It's undeniable that [*E.T.*] did give a big boost to the product" (Goodman-Malamuth 1985, 5).

Similarly, the comic strip *Archie* combined with General Foods Corporation to produce an "advertorial" about the Kool-Aid Man. In the summer of 1988, they spent "an estimated $25 million to boost sales of these packets of doctored sugar. On any given day, a 1 to 5 year old is as likely to drink Kool-Aid as orange juice" (Schmidt 1989, 6). Advertisers have even gone so far as to market "Coca-Cola clothes" to the junk-food crowd.

Pseudo-nutritionists often justify the massive amounts of sugar that Americans consume with the notion of our biological propensity to eat sweets. Supposed to have originally played the part in evolution that steered successful survivors away from unripe fruits and vegetables, this natural predilection for sweets is hyped to justify our present massive sugar consumption. One wonders how these defenders of the unhealthy status quo explain the French cuisine which has been resistant to the onslaught of even the sugar manufacturers. Certainly the power of advertising and the strength (or lack of strength) of cultural food traditions play a part. (I examine many of these issues of "culture" and food in chapter 6.)

Meat: Beef

Like sugar, today's beef is one of the least healthy foods. In its original place in Western civilization, it did provide many important nutrients—at least to the rich—its contemporary form is less than nutritionally desirable. Nevertheless, its profitability assures that the beef industry will continue to market it as a desirable food. And the fast-food industry has made beef the staple throughout its entire history.

Historically, meat has formed part of the human diet. As far back as one can trace, hunter/gatherers desired meat as a concentrated form of protein. Fiddes sees eating meat as evidence of man's (and woman's) rise to superiority over nature (1991, 2–3). The tribes that Marvin Harris (1985) studied in his defense of meat all crave it. Harris makes a logical case for even the rationality of cannibalism under certain conditions. "To believe that humans have no duties toward . . . the nonhuman world, has the implicit consequences of legitimating meat eating" (Fiddes 1991, 63).

Like sugar, beef has been the preserve of the rich. The meals of the agrarian elites were piled high with flesh. The desire to imitate the elite may be one reason that meat became so popular with the rest of society once it became more readily available.

For the agrarian poor, there was very little meat: "Before 1800, the average European ate about eight ounces of meat a year" ("Eschewing the Fat" 1991, 1). One will remember from our previous discussion that the basic European diet before sugar and tea was beer and bread—both made from grains. At this time, the obesity that excess meat consumption caused was seen as the desirable body shape. To be fat meant you ate well and therefore were rich and therefore were beautiful (see chapter 5).

Nutritionally, meat has the advantage of providing a complete protein. It provides ample amounts of vitamins B complex and E. Eaten raw it contains vitamin C, and it is the only plentiful source of B12—a problem for those vegans who do not take a dietary supplement. Meat also supplies trace minerals and trace metals not always easily obtained from other foods. But, even more than sugar, meat has come under attack by nutritionists for its lack of fiber, its fat and cholesterol content, and (lately) its embodiment of growth hormones, antibiotics, animal drugs, and the other chemicals from the grains cows eat.

In addition, what is often overlooked in the defense of beef is that the quality of the meat traditionally consumed was significantly different from that consumed today. Most meat was originally "game." Steers fed on open pastures with no pesticides sprayed on the inedible (for humans) grass that the steers ate. Steers also exercised. They therefore had a higher proportion of lean meat to fat. The cholesterol they did have was higher in HDL (High-Density-Lipo-Protein, the good kind) than LDL (Low-Density-Lipo-Protein, the bad kind).

People who ate beef before the agrochemical revolution were eating a significantly different meat.

In addition, the human style of life was significantly more active before the days of cars and power-operated equipment. People often walked miles to work. Their homes were much less well heated. Jobs often required much more muscle power.

We may assume that this created a situation where people developed larger arteries than those of our contemporary sedentary civilization. Like the Masai tribe of Africa, who drink high-fat cattle blood mixed with milk, diseases of the circulatory system did not develop. The plaque formed on the artery walls was not enough to cut circulation critically because of the large amount of exercise in their nomadic lifestyle (Bryant et al. 1985, 114–115). People died from other causes before clogged arteries became a problem.

Harris notes that beef was not originally America's favorite food. The Pilgrims liked pork. It was only with the conscious elimination of the Native Americans and their buffalo that America's Great Plains were made available for mass cattle herds (Harris 1985). At the beginning of the twentieth century, the railroad provided the means of transportation to central slaughtering in Chicago. Then, "the beef barons and packing-house owners—Armour, Swift, Cudahy, Morris—bought up the railroads, cornered the grain markets, and became as rich as modern-day oil sheiks" (Harris 1985, 118).

Responding to public pressure, President Wilson in 1917 sponsored a Federal Trade Commission investigation of concentration in the beef industry (Helmuth 1989, 3). As a result, in 1920 "the Attorney General filed a petition under the Sherman Act against the 'Beef Trust,' seeking to dissolve their monopoly" (Helmuth 1989, 3). A consent decree followed, which dissolved the Beef Trust.

Today, big business, mergers, and vertical integration act in consort with a technology of grain-raised feedlot cattle to produce a major part of America's oligopolized food industry. As Krebs writes: "cattle and meat packing rank as the largest single component of the food industry with regard to total assets, value added to product, total employees and total business receipts. With $33.8 billion in beef cattle sales in 1987 and with the meat marketing sales of $47.3 billion, it is the nation's fourth largest manufacturing industry" (1985, 2).

Once again, the meat industry has come to be dominated by three major packing companies: Iowa Beef Processors (IBP), Excel Corp (Cargill Corp's subsidiary), and ConAgra. They "now slaughter nearly 70% of all steers and heifers in the U.S." (Krebs 1990, 2).

Only with the development of the technology of feedlot cattle (no longer free grazing) and now "boxed beef" did beef consumption surpass pork in the mid-1950s (Krebs 1990, 23). And it was the success of McDonald's and the rest of the fast-food establishment that made beef the dominant American meat. While the rich could afford steaks, the middle class got the ground remains in hamburgers—initially at home and eventually in industrialized fast-food restaurants.

Between 1910 and 1976, American consumption of beef rose 72 percent (Brody 1987, 14). From 1976 to 1981, beef consumption declined by almost 20 percent to about 72 pounds per person per year. And it continues to decline, reaching 68 pounds in 1989 (Liebman 1990, 8).

Why the decline? Familiar to most readers is the concern with dietary fat causing the building up of cholesterol on the walls of blood vessels. This pervasive situation in America has made heart attacks and strokes our biggest killers (Brody 1987, 6). In spite of the massive efforts to advertise beef, consumers have been listening to the findings of scientific nutritionists. The power of science in an industrial age, which itself often changes its findings, has some standing amid the confusion and obfuscation in which the advertising industry has specialized.

More recently, nutritional science has been observing strong links between dietary fat and cancer. In its lead article for March 1991, *Consumer Reports Health Letter* begins with the headline "Does Meat Cause Cancer?" In more recent studies, the newsletter reports, "Higher intake of saturated fat was associated with increased risk of colorectal [colon and rectal] cancer: dairy products and meat figured equally in that increase" ("Does Meat Cause Cancer?" 1991, 2). Marilyn Kaggen reports that the *only* serious way to decrease the risks of breast cancer is to lower fat intake (1992, 66–68).

Robbins writes that factory farmed meat has thirty times more saturated fat than naturally raised meat (Crawford 1975, 24). And because eating meat is "high on the food chain," one encounters a lot of the chemicals in the grown grain as well as those placed specifically in the animal feed.

The relatively new organic health food supermarkets in Boston advertise that beef contains about 200 chemicals. The biggest areas of concern are the petrochemicals in the grain (pesticides, herbicides, and fungicides), the antibiotics, and the growth hormones added to the grain fed to the cows. David Pimental of Cornell University reports that "in the last 40 years, the volume of pesticides used in American agriculture has increased tenfold" (Montgomery 1987, 7). Because pesticides concentrate in the fat of the animal (and in the liver), one should be even more concerned with the consumption of nonorganically raised animal products. Robbins reports that eating meat has fourteen times more pesticides than plant food.

In a profit system, the cheapest price "succeeds." In this context, large producers have found that they can stimulate a cow's growth and reduce or eliminate illness by using growth hormones and subtherapeutic doses of antibiotics. Because of growth irregularities, the European Economic Community (EEC) has banned all hormonal growth-promoting drugs except for therapeutic uses. It is generally agreed by the medical profession that by feeding animal antibiotics, we are beginning to lose the effectiveness of some of these antibiotics in the same way that overprescription of them restricts their effectiveness. Specifically mentioned are penicillin and tetracycline. Over half the antibiotics manufactured are now sold to animal producers.

Furthermore, the role of government is less than satisfactory in monitoring beef. *Nutritional Action Health Letter* reports that the Food Safety and Inspection Service of the U.S. Department of Agriculture "doesn't even routinely test for about 80% of the pesticides that could be present" (1989, 6). In one instance, "routine testing was dropped of 11 meat producers who agreed to take extra precautions against illegal residues. Eight of the 11 never fully complied" (1989, 6). "In fact, the U.S.D.A. [U.S. Department of Agriculture] tests only *one out of every quarter million* slaughtered animals for toxic chemical residues" (Robbins 1987, 338, citing *Mainstream* 1983, 17). Once again the U.S. government, by not testing thoroughly, tends to favor its accumulation function (facilitating profit making for the beef industry) over its legitimation function (making beef safe for the consumer).

Nutritional Self Defense: Vegetarianism

If in the food supply meat (and especially beef) is unhealthy, then what steps should Americans take to defend themselves? This is not an easy question because, as we discover in chapter 6, our tastes are flexible and can be significantly influenced by advertising as well as by the cultural traditions in which we grew up.

One approach is to alleviate all the confusion and contamination worry by throwing up your hands and eating whatever you want with abandon. Although the death and disease figures from food adulterants are only percentages, the risks are real. Only the cavalier ignore all the warnings from nutritional science in the past twenty years.

Another response is to cut down on red meat consumption and switch to other meats. But, contamination is often just as real with these other meats (see chapter 1 on chicken). By reducing meat consumption in general, you lessen the amount of contamination. This is the solution advocated by many who have studied the Chinese diet. They are recommending that meat be used as a supplement or condiment to a main diet of grains, fruit, vegetables, and dairy.

In addition, it is possible to find selected producers of beef without any growth stimulants or antibiotics that graze rather than feed in lots. There are some producers of organic beef. If one "needed" beef in one's diet, this might be a rational solution.

An increasing number of people have come to regard some form of vegetarianism as desirable. Mauer (1989) estimated them to be 7 million (see chapter 10). People categorize vegetarians differently. A summary classification, going from the least to the most pure with all of them eating fruits, vegetables, and grains, is shown in Table 4.1.

Table 4.1 Types of Vegetarians

Type	Meat	Other
Semivegetarians	chicken, fish (no beef)	dairy, fruits, vegetables, grains
Pollo Vegetarians	chicken (no beef)	dairy, fruits, vegetables, grains
Pesco Vegetarians	fish (no beef)	dairy, fruits, vegetables, grains
Lacto-Ovo Vegetarians		milk, eggs, fruits, grains, vegetables
Lacto Vegetarians		milk, fruits, vegetables, grains
Ovo Vegetarians		eggs, fruits, vegetables, grains
Vegans		grains, fruits, vegetables

There is no question that even a vegan diet can be adequate nutrition for most adults (Mudambi and Rajagopal 1987). You need to know how to balance proteins. But if you stay within the traditional food combinations (such as rice and beans or corn and beans), you will get enough protein. Recently, nutritionists in the United States have started noting that Americans eat so much protein that unless a person is a strict vegan, he or she would have a difficult time being protein deficient. The lack of vitamin B12 can be solved for vegans by nutritional yeast or a vitamin supplement.

Of course you should not jump directly from a meat-based diet to that of a vegan, except in extreme cases. Mauer (1989) and Whit (1990) have identified a *process* of progressive movement toward a purer vegetarianism. Different people stop at different points but may progress further at a later time.[2]

In this ideal-typical progression, most people start with some concern for health. These may raise questions about red meat's role in arteriosclerosis, cancer, and digestive problems (lack of fiber), which may lead a physician to order a patient to cut red meat from the diet. The process then becomes more interesting. Generally we accept values that are consistent with our practice. When we stop eating red meat, we can identify with others in the same situation. We can think of ourselves as "vegetarians" of the mildest sort. This new self-conception gives a greater acceptability to arguments that mention the numerous reasons for abstaining from red meat. These reasons may lead involvement with "significant others" (close friends, role models) who have progressed further on the vegetarian continuum toward the position of vegan. Because our ideal-typical person has already achieved some benefits from ceasing the eating of red meat, she or he may be inclined to consider a further reduction in flesh in the diet. Usually this involves the elimination of chicken and fish.

A person can progress further toward the vegan goal. She or he may associate with some vegans, join a health food co-op, or begin reading publications that champion vegetarianism. In addition, if someone feels better after making these changes, there is more incentive to continue the process.

As you move along the continuum, more and more of the reasons for becoming a vegetarian can be included in your value system. Whereas, initially, you may have begun the process purely for better health, you may soon find that the arguments about solving world hunger and animal cruelty make sense. You may even become politically active in vegetarianism as a political movement. Often these activities have religious overtones: "Modern vegetarianism includes a documented set of beliefs that look remarkably religious. It involves the expression of views and ideology about the relationship of humankind to the world in which we live . . . and the path to salvation to be followed" (Murcott 1986, 12). Nutritional anthropologists Kandel and Pelto have termed this a "movement of social revitalization or transformation . . . another realm common to religion" (1980, 332).

To understand vegetarianism fully, it is necessary to view the multitude of reasons why people become vegetarians. You, the reader, should keep in mind that various reasons can apply at various stages in the process of becoming a vegetarian.

You can not only improve ill health by eating less meat; you can also excel physically and spiritually. Of the latter, Hinduism and Jainism in India are advocates. Of the former, there is good evidence that superior athletes can be bred on a vegetarian diet. Jane Brody notes experiments back in 1904 where vegetarians exceeded nonvegetarians by almost double on a test of squeezing a grip meter (1987, 442). In addition, such famous athletes as Paavo Nurmi (twenty world running records), Bill Pickering (British English Channel swim record breaker), Murray Rose (triple gold medal in Olympic swimming), and Bill Walton (basketball star of the Portland Trailblazers) were all vegetarians (Brody 1987, 443). Even football teams are now eating pasta (complex carbohydrates) instead of loading up on steak as they did twenty years ago.

Another major consideration is world hunger. Because it takes 9–16 pounds of grain to produce a pound of beef, meat is the most wasteful of protein resources. As John Robbins writes: "By cycling our grain through livestock . . . we . . . waste 96% of its calories, 100% of its fiber and 100% of its carbohydrates" (1987, 3). "The world's cattle alone . . . consume a quantity of food equal to the caloric need of 8.7 billion people—nearly double the entire population of the planet" (Robbins 1987, citing Rensberger 1974, 14). If the entire world ate a vegan diet, there would be enough food produced to provide all of us with a diet that would make us fat—and a healthier kind of fat too! Unfortunately, in the world economic system, food is sold to the highest financial bidder. It is not uncommon for developing countries with starving people to export food. It has happened in the Sudan and throughout Africa. "In Guatemala 75% of the children under five years of age are undernourished. Yet every year Guatemala exports 40 million pounds of meat to the United States" (Robbins 1987, 352; and see chapter 9).

A fourth complex of reasons for becoming a vegetarian relates to ecological damage to the planet and to nature. Ecology has recently become much more popular. Beef (and meat) production causes documented harm to a number of different areas of the earth. Water is becoming increasingly scarce in a number of areas of the United States. Robbins notes that water is needed both to grow the food for beef and to wash away their excrement: "To produce a single pound of meat takes an average of 2,500 gallons of water—as much as a typical family uses for all its combined household purposes in a month" (Borgstrom 1981). While a day's food for a meat eater requires 4,000 gallons of water, that for a vegan requires 300 gallons. "It takes less water to produce a *year's* food for a pure vegetarian than to produce a *month's* for a meat-eater" (Altschul 1965).

The meat industry receives enormous subsidies by means of the provision of water in the country: "If the cost of water needed to produce a pound of meat were not subsidized, the cheapest hamburger meat would cost more than $35 a pound!" (Robbins 1967, 367).

Water devoted to beef feed could be used for generating energy so that the Pacific Northwest would not require two nuclear plants. And California, already short on water, is the Northwest's biggest water consumer (Robbins 1987, 369). Too often, this type of irrigation causes environmental damage by salinization and heavy metal pollution.

The other problem with water is its contamination by animal excrement. Because animals in feedlots produce so much in one place, the groundwater often gets contaminated. "One cow produces as much waste as 16 humans. With 20,000 animals in our pens, we have a problem equal to a city of 320,000 people" (Robbins 1987, 372, citing Harry J. Webb). And "the meat industry single-handedly accounts for more than three times as much harmful organic waste water pollution as the rest of the nation's industries combined" (Borgstrom 1981)!

The depletion of the world's forests is another reason vegetarians cite for not eating meat. Once again, profitability is the key. In the developing world, it is well known that the potential medicai and nutritive value of rain forests is being lost to grazing land for beef. Costa Rica, in Central America, is but one of many countries where this is a common practice. Because the owners of beef herds can make more money razing forest for grazing land, they choose to grow and export this beef to the American fast-food industry. This takes land away from substantive production and allows more exploitation of workers.

In addition, American forests are cut for grazing land: "For each acre of American forest that is cleared to make room for parking lots, roads, houses, shopping centers, etc., seven acres of forest are converted into land for grazing livestock and/or growing livestock feed" (Robbins 1987, 360, citing Harand Fielde 1984).

There are concerns that the methane gas the cattle excrete is contributing to the thinning of the ozone layer in the atmosphere. The world's 1–3 billion cattle annually release 70 million tons of methane gas. This is one-fifth of the green-house gasses.

People are upset with the industrialized killing of cattle. The older method was to hit them on the forehead (the man standing above) with a poleax (something like a sledgeham-mer). The problem is that, because the cattle are frightened, their heads are in motion. Thus, even the best worker often misses. The result is the agony of the animal's being hit a number of times before it is beaten to death. Modern methods employ a stun gun. But it often fails to kill cows instantly. Once again the animal must be beaten to death or have its throat cut.

Not all of the animals die immediately. With less than a minute devoted to each animal, it is not uncommon to hook them up to the ceiling by means of their rear leg (tearing skin, ligaments, joints and muscles) while they are still alive and struggling. Animals killed for kosher beef suffer even more in the absence of the stun gun and pole ax. They must suffer up to five minutes hanging upside down (Singer 1975, 154–155). If people knew the killing process and the conditions of labor in America's industrialized slaughterhouses, there quite pos-sibly would be many more vegetarians.

Concerns have also been raised about the treatment of animals on the way to slaughter and in the process of raising them. Tales are common of cattle crowded together in box-cars, sometimes injured or trampled by other frightened cattle. Most outrageous are the stories of raising veal. Calves, at birth, are taken immediately from their mothers, put in small stalls, and fed an unnatural liquid diet without iron (forcing them to become anemic). They are kept in complete darkness (except for two daily feedings), and many go completely blind. They are fed chemicals so that when they reach 350–400 pounds they can be slaughtered. All this for the "taste" of veal eaters. British sociologist Ann Murcott sums up the position of vegetarians from a social-psychological perspective, "When [people choose] to avoid eating flesh . . . they are making statements about themselves, answering the unspoken question of 'who am I' by saying that they are the right thinking sort of a person with a concern for their health, the state of the world in which they want to live and the sort of qualities they regard as essentially human" (1986, 13).

Summary

This chapter has covered the manner in which sugar and meat are overproduced in a society in which the major concern is profit. Sugar is added to a wide variety of products. It pervades soft drinks and cereals. And, because there is only a limited number of cereal and soft-drink producers, there is little possibility of serious competition from a better product. That is the nature of oligopoly capitalism. The production of beef is increasingly less competitive and, once again because of the profit motive, effects on health, the environment, and animals are ignored.

Vegetarians have made a response to the problem at the level of consumption. And they may form an eventual resource for political actions that could be ecologically beneficial. But, generally, the positions of the rich and the poor (and middle class) through history have been reversed. The rich once ate much meat and were obese, but it was better meat. And they presumably exercised. The poor have traditionally been the eaters of grain and, albeit involuntarily, almost vegetarians. Yet their traditional problem has been undernutrition. Worse, with the substitution of sugar and tea for bread and beer in England, the poor had their health significantly endangered, while the rich benefitted from the new sources of profit.

Today the situation is reversed. The rich (and some of the middle class) have used their superior education to listen to nutritional science. The rich eat a lot less beef and sugar. Salads, low fat, and low calories are the rule. As the contemporary American saying goes, "You can never be too rich or too thin." The poor and middle class, in their rush to emulate the rich of previous generations, have been manipulated into the most ecologically destructive yet most profitable kinds of consumption. Fast-food meat provides an enormous profit for the few multinational corporations that control its disposal. Not content with just America, McDonald's and others have successfully opened markets in Europe, China, India and the rest of the developed and developing world.

One health problem resulting from fast-food and convenience-food operations is obesity. Because the poor of the world are much more influenced by advertising than the rich, they succumb to it (and emulation of the rich) in accepting the fat-causing food America is producing. The number of overweight people continues to grow.

Americans are caught between the media norms of thin female beauty and the economic imperative of increasing consumption of fat-causing, highly processed food. Chapter 5 explores this problematic.

References

"Americans for Safe Food." 1985. In *Guess What's Coming to Dinner*, ed. Ben Mckelway. Washington, D.C.: Center for Science in the Public Interest.

Acres, U.S.A. 1985. Kansas City, Mo., June, 2.

Altschul, Aaron. 1965. *Proteins: Their Chemistry and Politics*. New York: Basic Books.

Bavan, Paul, and Paul Sweezy. 1966. *Monopoly Capital*. New York: Monthly Review.

Belasco, Warren. 1989. *Appetite for Change*. New York: Pantheon Books.

Bogstrom, Georg. 1981. Presentation at the annual meeting of the American Association for the Advancement of Science.

Brody, Jane. 1987. *Jane Brody's Nutrition Book.* New York: Bantam Books.

Bryant, Carl A., Anita Courtney, Barbara A. Markesberry, and Kathleen M. DeWalt. 1985. *The Cultural Feast.* New York: West.

Crawford, M.A. 1975. "A Re-evaluation of the Nutrient Role of Animal Products." In *Proceedings of the Third World Conference on Animal Production*, ed. Edward Reid. Sydney, Australia: Sydney University Press.

Dixon, Jennifer. 1992. "Meat Inspection Agency is a Failure." *Grand Rapids Press*, March 29, D1.

"Does Meat Cause Cancer?" 1991. *Consumer Reports Health Letter*, March, 1.

Douglas, Constance, and William Hefferman. 1989. "IBP's Dominance in the Meat Packaging Industry." Paper presented at meeting of the Food, Agriculture, and Human Values Society, Little Rock, Arkansas.

Dufty, William. 1975. *Sugar Blues.* New York: Warner Books.

Durning, Alan B. 1991. "Fat of the Land." *World Watch*, May/June, 11–17.

"Eschewing the Fat." 1991. *Harvard Health Letter*, March, 1.

Fiddes, Nick. 1991. *Meat: A Natural Symbol.* London: Routledge.

Forman, Monte, and Marjorie Forman. 1990. *Fast Foods: Eating In and Eating Out.* Mt.Vernon, N.Y.: Consumer's Union.

Harris, Marvin. 1985. *Good to Eat.* New York: Simon and Schuster.

Helmuth, John W. 1989. "Meat Packer Concentration." Address presented to the Dakota Resource Council annual meeting, Mandan, North Dakota, November 4.

Hur, Bobin, and Davis Fields. 1984. "Are High Fat Diets Killing Our Forests?" *Vegetarian Times*, February.

"Is There Still Anything Safe to Eat?" 1988. *Tufts University Diet and Nutrition Letter*, August, 3–6.

Jacobson, Michael F., and Sara Fritschner. 1986. *The Fast-Food Guide.* New York: Workman.

Jerome, Norge. 1980. "Nutritional Anthropology." In *Nutritional Anthropology*, eds. Norge Jerome, R.F. Kandel, and G.H. Pelto. Pleasantville, N.Y.: Redgrave.

Kaggen, Marilyn. 1992. "Breast Cancer: Wishes, Lies, and Profits." Z, May, 66–68.

Kandel, R.F., and G.H. Pelto. 1980. "The Health Food Movement: Social Revitalization or Alternative Health Maintenance System." In *Nutritional Anthropology*, ed, Norge Jerome, R.F. Kandel, and G.H. Pelto, 327–363. Pleasantville, N.Y.: Redgrove.

Knapp, Arthur W. 1920. *Cocoa and Chocolate: Their History from Plantation to Consumer.* London: Chapman and Hall.

Krebs, A.B. 1990. *Heading Toward the Last Roundup: The Big Three's Prime Cut*. Washington, D.C.: Corporate Agribusiness Project.

Lefferts, Lisa. 1989. "Pass the Pesticides." *Nutrition Action Health Letter*, April, 5–7.

Lichtenstein, Alex. 1991. "Sweet Misery: Sugar Creates Slave Labor in Florida's Fields." *In These Times*, January, 16–22.

Liebman, Bonnie. 1985. "Meat Considered." *Nutrition Action Health Letter*, May, 4–7.

———. 1986. "Kids and Diet." *Nutrition Action Health Letter*, April, 1, 6–7.

McKeown, Thomas. 1976. *The Modern Rise of Population*. New York: Academic Press.

Mauer, Donna. 1989. "Becoming a Vegetarian: Learning a Food Practice and Philosophy." Paper presented at the meeting of the Association for the Study of Food and Society, College Station, Texas, June.

Mintz, Sydney. 1985. *Sweetness and Power*. New York: Viking Penguin.

Montgomery, Ann. 1987. "America's Pesticide-Permeated Food." *Nutrition Action Health Letter*, June, 1, 4–7.

Mudambi, Sumati R., and M.V. Rajagopal. 1987. "Is a Vegetarian Patient at Risk?" *Nutrition 3*(6):373–378.

Murcott, Anne. 1986. "You Are What You Eat: Exploration of Anthropological Issues Influencing Food Choice." In *The Food Consumer*, ed. George Ritson. Chichester, England: Wiley.

Rensenberger, Boyce. 1974. "World Food Crisis: Basic Ways of Life Upheavals from Chronic Shortages." *New York Times*, November 5, 14.

Rifkin, Jeremy. 1992. "The Last Roundup for Beef?" *Countryside Magazine*, May, 84–86.

Robbins, John. 1987. *Diet for a New America*. Walpole, N.H.: Stillpoint Publishing.

Schmidt, Stephen B. 1989. "Hawking Food to Kids." *Nutrition Action Health Letter*, January/February, 5–7.

Singer, Peter. 1975. *Animal Liberation*. New York: Avon Books.

"Simply Sugar or More Than Sugar?" 1988. *Tufts University Diet and Nutrition Newsletter*, July, 3.

Tannahill, Reay. 1973. *Food In History*. New York: Stein and Day.

Webb, Harry J. Blaine, Nebraska. n.d.

Whit, William C. 1990. "The Meaning of the Health Food Movement." Paper presented at the World Congress of Sociology, Madrid, Spain.

"Who's Checking the Beef?" 1989. *Nutrition Action Health Letter*, April, 6.

Notes

[1] 1.Due, in part, to a blight on coffee beans, the British Empire sponsored sugar production in the West, and tea production in the East India Company. Both tea and sugar lacked substantial nutrients. They became the first high-profit junk foods generally available in the industrializing world. Because they could be priced cheaper than homemade beer and bread, they accounted for an increasingly large share of the working-class diet (along with potatoes—the new crop imported from America). But tea and sugar accounted for some of the malnutrition of the working class, especially women and children. In essence, sugar and tea performed the function of population control through a higher infant mortality from malnutrition. Though Mintz (1985) argues for a reduction in total working-class nutrition, McKeoun (1976) argues that better farming methods and the potato more than make up the difference.

[2] 2.Health and vegetarian social movements are discussed in chapter 10.

Chapter 5

Fat and Thin: Obesity and Anorexia

> Girls try so hard to conform they'll even puke to belong,
> Then 3 years later they have false tits sewn on.
>
> — Brenda Kahn, 1990

Weight is perhaps the most important issue for American women. Of concern are elements of physiology, biology, nutrition, junk food, romance, character, stigmatization and power. Women are caught in a fundamentally contradictory social position in which they are told both to eat the manifold fattening food products produced for profit in our society and to stay pencil thin. The eating disorder bulimia is a physical expression of this contradiction. Binging and purging become a response that mirrors the opposed pressures under which women live.

This chapter begins with some reflections on the role of fat and thin in history. It then examines contemporary forms of fatness and thinness such as obesity and anorexia. It looks to some contemporary feminist scholarship for insight into the relationship between social structure and body size problems.

Fat and Thin in History

Physiologically, fat is a mechanism for storing energy. One may suppose, in past centuries, that fat got people through times when there was not enough to eat. When the body cannot feed from the world around it, it feeds on its fat reserves. Before antibiotics, many diseases (e.g., tuberculosis) required substantial fat reserves for sustenance during illness.

Fat operates in distinctive ways for women. Robbins writes, "The principal trigger for the onset and maintenance of regular ovulatory cycles in human females is a critical layer of body fat." "From a male perspective, fat would be a desirable characteristic to look for in a mate because it would be a sign that she was likely to be fertile" (Robbins 1989, 9). Some people believe that because a woman's body can feed off its fat resources better, women (with equal training) would make better long distance runners.

Before the Industrial Revolution, fatness was a mark of status. It meant that you did not need to exercise as much as the laboring lower classes. If you were rotund, you were eating well. Throughout the Middle Ages in the Western world, obesity for the rich coexisted with slenderness for the peasants and laboring classes. The exception was *anorexia mirabilis* (Brumberg 1988, 45). Certain women denied themselves physical nourishment in return for spiritual nourishment: "Through fasting, the medieval ascetic strove for perfection in the eyes of her God" (Brumberg 1988, 45). Christianity contained, from its beginning, an ascetic tradition of food denial. Even Jesus fasted as a route to holiness.

By the 1800s medical science began to provide an alternative context in which to interpret food avoidance. Brumberg provides examples of the attempts to document scientifically the total food avoidance of these "fasting girls" (1988, 65). By about 1900, treatment of this kind of extreme food avoidance was medicalized and is now called *anorexia nervosa*, a modern eating disorder.

Women's lot throughout most of history was to manifest fatness. It was (and is still in most of the developing world) a standard of beauty. What evidence we have would declare that such beauties as Cleopatra and the nudes who posed for Titian, Rubens, and Renoir are, by contemporary standards, significantly obese. Yet their fat demonstrated their affluence. The rich usually set an era's standards of beauty.

In the early 1900s, the life insurance companies discovered that mortality was in part tied to obesity. Here began the "height to weight charts" that now form the basis for insurance company mortality predictions and our norm (albeit sometimes changing) of normal, obese, and too thin.

During World War I, rationing necessitated meatless days and the use of "skim milk, molasses and margarine" (Schwartz 1986, 141). The Hoover administration asked people to eat "more fruits, vegetables, and corn (or rye, or graham) flour, less meat, fat and sugar" (Brumberg 1988, 141). Note the remarkable resemblance to the contemporary government recommendations about improving diet by the McGovern Committee (see chapter 3).

It was also in the Hoover era when the style for female body image made the switch from fat to thin. The "flappers" of the 1920s pictured beauty as boyish. Women were to be thin and to have small breasts. Women often wrapped their breasts to conform to that image. The thin image corresponded to the increasingly politically active role of women. In 1921 the political agitation of a decade resulted in women's winning the right to vote. Parallel to that political advance was a body image with short skirts and smoking (itself a method of weight reduction), which signaled female cultural "liberation."

The 1929 Depression brought a concern with health food, dieting, and calorie counting. It is reported that Henry Ford followed the Hay diet, eating only one type of food at a meal. Ford ate, "only fruits at breakfast, only starches at lunch, and only proteins at dinner" (Schwartz 1986, 200). The ideal weight was in decline.

Thin and Fat in Contemporary Society

For the rest of contemporary history, thin has been *in*. Although desirable hairstyles, leg lengths, and breast size have altered, the one constant is thinness. This makes that most popular woman beauty, Marilyn Monroe, a bit of a "cow" by contemporary standards.

But thinness is not available to everyone. A very small percentage of women manifest the superslender silhouette of the 1960s model Twiggy. Therefore thinness, like fatness in previous eras, becomes a mark of the elite. "You can never be too rich or too thin." If few women have the natural bodies to attain the ideal, the affluent women of leisure can devote adequate time to thinning, while the rest of the population must do the (mostly sedentary) jobs of advanced industrial society and eat the food provided therefrom.

Nevertheless, the middle classes generally emulate the rich. Just as peasants wished to be fat in previous eras, now middle-class women wish to be thin. The commercial media push thinness. Most of the models and actresses in the media are thin. Joan Jacobs Brumberg inserts this phenomenon into the famous *The Theory of the Leisure Class* (Thorstein Veblen 1922). Brumberg writes that "a thin, frail woman was a symbol of status and an object of beauty precisely because she was unfit for productive (or reproductive) work. . . . According to Veblen, a thin woman signified the 'idle idyll of the leisure class'" (1988, 185).

Class and gender intersect to move toward cultural norms to which only the leisurely (rich) have time to adhere. Class and gender differences are part of the basic structure of a capitalist society. One would expect that capitalists would profit from (and therefore help perpetuate) the cultural norm of slenderness. Hillel Schwartz in his intriguing *Never Satisfied* performs a brilliant analysis of the manner in which this dynamic is played out:

> It happens in the diet industry where consumers spend $30 billion dollars a year to get (and stay) thin. Yet it never quite works. . . .
>
> Dieters ultimately consume more. The diet is a supreme form for the manipulation of desire pre-

> cisely because it is so frustrating. Capitalists have a vital stake in the failures of dieters as in the promotion of dieting. It is through the constant frustration of desire that Late Capitalism can prompt even higher levels of consumption. . . .
>
> Dieters, like infants, give the food industry the opportunity to profit from chaff, tailings and waste. Beef tallow and the by-product vegetable oils go into margarine and then into lighter "spreads." Bran and wheat germ stripped from grain during milling, become expensive food supplements for dieters who must assure their health as they eat less. Skim milk, whey and milk solids go into powders for liquid diets. . . . Drug companies find markets for glandular substances that would otherwise be discarded by slaughterhouses. With half a cent worth of vitamins . . . a manufacturer can make cellulose, soybeans, and spices into a vitamin-rich diet food (FN 20).
>
> Dieters must constantly change wardrobes as their shapes and weights go in and out, up and down. Dieters must always seek out new diet foods, new diet books, new devices . . . the light foods, more synthetic, less substantial, give a greater return on the dollar. . . .
>
> An expanding Late Capitalism requires that no one ever be fully satisfied. (1986, 328–329)

In the context of enforced thinness, life is not easy for contemporary fat people. From a sociological perspective fat serves the same stigmatization function that skin color and sex serve in theories about the stigmatization of minority groups.

Jeff Sobal of Cornell University has summarized the work of Natalie Allon in the 1970s, a forerunner of food and society analyses. As a mark of minority-group status, Allon noted that in overweight adolescents, ". . . fatness serves as the exclusive focus of interaction, being an overriding quality of the overweight girls' identity" (Sobal 1984, 17). Fatness influenced and changed other people's normal feelings about people, biasing them in a negative manner: "being obese degraded virtues which were not related to weight . . . [resulting in] . . . narrowly qualified complements which focused upon isolated attributes and ignored their holistic identity . . . " (Sobal 1984, 17). A contemporary typical example includes the common "You have such a pretty face."

Contemporary analysts report consistent findings. Meadow and Weis write that "fat oppression . . . remains one of the few 'acceptable' prejudices still held by progressive persons" (1992, 133). Jack Levin (1988) notes that "fat people frequently receive contempt rather than compassion unless their obesity can be attributed to some physical ailment (for example, 'glandular condition')" (1988, 61). Like other minority groups, fatness "is used to stigmatize an entire group of human beings by making their belt size an excuse for prejudice.-.... They frequently have trouble getting married, going to college, obtaining credit from a bank or being promoted. In short, they are excluded, exploited and oppressed" (Levin 1988, 61). In reaction to the stigma attached to fatness, Allon found that overweight adolescents blame themselves (Sobal 1984). Like many African Americans and gay people, fat people often internalize the way the external world treats them. This laying of blame on the individual allows all the responsibility for obesity to fall on the individual rather than on a society that profits from food that makes one fat (as well as the means to take fat off).

Sociologists refer to this phenomenon as "blaming the victim." Rather than perform an analysis of the *social* forces that lead to obesity, the analysis is pitched at the level of psychological deficiency. Obese people "lack self control." They eat to replace the love they lacked as children. They eat because they are afraid of sex or success or intimacy. They eat to insulate themselves, eat to replace the children who have left home (mothers), or eat to express anger. They are narcissistic. They are anti-male. They are depressed. Orbach writes: "Since the Second World War, psychiatry has . . . told unhappy women that their discontent represents an inability to resolve the 'Oedipal constellation'" (1978, 4–5). Once again, psychiatry provides a legitimation of the status quo.

Usually, little attention is given in this list of victim blaming "analyses" to the manner in which, in a capitalist society, profit is generated from processing food. Therefore, the heavily advertised food products are usually those that are the most processed and fattening. Though nutritional science may proclaim contrary imperatives, dietitians and nutritionists do not command the financial power to use the best psychology available to sell healthy products in America. The advertising budget of Coca-Cola alone would support more adequate nutritional research than the budgets of all the legitimate scientific nutritional publications in the United States combined.

In response to the denigrations and self-hate involved in victim blaming, an alternative analysis has evolved from American feminism. Its leaders are Susie Orbach (1978), Kim Chernin (1981a, 1981b, 1985), Hilda Bruch (1979), and Naomi Wolf (1991).

Unlike the victim blamers, feminist analysts see obesity as a response to the oppression of women in contemporary society. Unlike "reaction" (which is automatic), "response" implies some degree of choice and potential control over one's actions. It avoids both the "person as automton" model that has characterized some of liberal sociology ("mechanical Marxism") and the overly voluntaristic ("people are totally free to choose") positions which conservative thinkers are prone to espouse.

Chernin and Orbach begin from the contemporary situation in which men and women are treated differently. Men *are*, women *appear*. Because appearance is so important to the usual treatments of women, fat stigmatizes and disadvantages them.

Orbach is from a psychological/analytic background. She makes the assumption, common to most analytical and counselling psychological theory, that she interprets correctly the subconscious or unconscious of fat women. Of course, the rest of the nonfeminist psychological establishment holds different theories based, usually, on less social and more "proper" psychological factors (inadequate socialization, rewards for eating, incomplete selves).

Chernin takes the position that "at the heart of an eating obsession is a regression to the infantile condition" (Chernin 1985, 197). But she gives it a psychoanalytical, feminist twist: "Hidden behind our obsessive preoccupation with food is a need to regain a relationship to the sacral mystery of female being. . ." (p. 197). Chernin tackles the issue of childhood refusal to eat (common in most homes in one form or another) with an almost metaphysical grounding in food: "Food is so charged, so significant, so informed with primal meaning and first impressions of life, mothering, and the world that we might well expect communications to take place through food to carry even more weight than those that arise when a child totters about . . ." (p. 103).

Susie Orbach is more societally grounded. She says that her project is "always to see the social dimensions which have led women to choose compulsive eating as an adaptation to sexist pressure in contemporary society" (1978, xvi). Like "response," "adaptation" connotes some individual choice in confronting social conditions. Orbach believes the victim-blaming psychological approaches to be a serious distortion of the reality she perceives. Instead, she portrays the "central issues of compulsive eating [as] . . . rooted in the social inequality of women" (1978, 5).

Because the *appearance* of women is so much emphasized in a society that sells products through the norm of the beautiful (i.e., thin) woman, the vast majority of the female population are continuously dissatisfied with themselves. This position leaves them open to manipulation by sponsors of corporate attempts to reshape their bodies (the diet and exercise industries). Because they are influenced to conform to an *external* standard, they often feel victimized and out of control.

Compulsive eating is one response (among others) women make to this victimization: "For many women, compulsive eating and being fat have become one way to avoid being marketed or seen as the ideal woman" (Orbach 1978, 9). This socially based psychological/biological response "expresses a rebellion against the powerlessness of the woman" (Orbach 1978, 9). Wives and mothers in their traditional roles are expected to be the givers in the family: "I feel empty with all this giving so I eat to fill up the spaces and give sustenance to go on giving to the world" (Orbach 1978, 12).

Fat allows women to be taken seriously. Unfortunately in our "civilized" society, power is still often tied to size. And large size can be a means of power: "When I'm fat I feel I can hold my own. Whenever I get thin, I feel I'm being treated like a little doll" (Orbach 1978, 13).

As fat persons, women are usually desexualized. One will therefore not confuse a man's sales pitch with a romantic come-on. And people will not accuse fat women of "coming on" either.

Like victim-blaming psychologists, Orbach spotlights the mother-daughter relationship. But she focuses on the ambivalence that a mother feels about socializing a daughter (often with food) into a world where female roles are to be inferior and mostly giving. She notes great ambivalence around feeding. Mothers want to give to their daughters. But daughters have to face a world where men come first. In many households, men are fed either first or better or both. Female children should not expect unconditional giving. And with food, mothers must be careful not to give too much because of the danger of obesity. Daughters who consistently eat too much will go into a thin-oriented world at a distinct disadvantage. Discipline must be learned even as a child.

Daughters also have ambivalence about their mothers that can be expressed through food. In a bit of "psychologeeze," which Orbach claims evidences the ambiguities daughters feel toward their mothers, she writes: "In overfeeding herself, the daughter may be trying to reject her mother's role while at the same time reproaching her mother for inadequate nurturing" (1978, 21).

Thus, from Orbach's sociological/feminist perspective, fat becomes a *response* to inequality and oppression whether in the society at large or in the family of origin as it reflects the outside society: "Fat is a way of saying 'no' to powerlessness and self-denial" (Orbach 1978, 21).

A Multidisciplinary Analysis of Obesity and its Therapies

There is so little authoritative analysis of the causes of obesity that one might take the type of analysis employed as more an evidence of the scholarly background of the analyst than of "scientific" discipline. Of general import is Spitzack's insight that "historically there exists a connection between insufficient knowledge within the medical community and the attribution of causality to the individual" (1990, 29).

We do know that humans have some natural taste for sweets. And we assume that it was an evolutionary endowment that enabled our forerunners to eat ripe fruits instead of rotten or unripe ones. Yet the creation of the nutritional monstrosities that advanced capitalist industrial societies offer through their vertically integrated supermarket chains and convenience stores is hardly the evolutionary fulfillment of biological drives. In fact, from the point of view of contemporary nutrition, our biological drives may be a handicap in a world filled with sweetened and nutritionally unhealthy snacks and meals.

There are explanations other than the psychological and sociological ones already mentioned. These explanations include heredity, metabolism, fat cell theory, satiety and hunger theories, learning theory, and body-image distortion theories. Among the biologically based theories, heredity and metabolism predominate. Generally, if both parents are obese, one can expect at least a tendency (determinists would use stronger language) toward obesity in their children. A large part of obesity is hereditary.

Among metabolic theories, endocrinological and set-point theory predominate. As Sobal and Muncie write: "Endocrinologic disturbances such as thyroid, adrenal, or hypothalamic disorders may lead to obesity" (1990, 1241). Of more general importance, set-point theory postulates that a "set point for a person's body weight is influenced by genetics and environmental factors, and can be reset by diet, exercise, and medications" (Sobal and Muncie 1990, 1241).

No longer in vogue, fat-cell theory postulated critical growth periods at which fat cells are formed. But we now know that these fat cells develop at all ages (Sobal and Muncie 1990, 1241). In addition, "satiety and hunger theories focus on somatic cues for eating and stopping eating, with problems in perception of cues and feedback rates leading to overeating" (Sobal and Muncie 1990, 1241).

Among the analyses that deal with socioeconomic status, Sobal and Stankard have demonstrated a significant relationship between low socioeconomic status and obesity (1989, 160–275). While one "can never be too rich or too thin," obesity is more common among poorer people. It also may be found desirable by ethnic and racial groups who either come from cultures that value obesity (most of the developing world) or who have a predominance of people of lower socioeconomic status in their group.

In the process of making profits, the general condition of American women who perceive themselves as overweight fosters a health and diet industry. Naomi Wolf documents "the $33 billion-a-year diet industry, the $20 billion cosmetics industry, and the $300 million cosmetic surgery industry" (1991, 17). Buying beauty products has replaced the suburban housewife's spending on furnishings (Wolf 1991, 66).

The Associated Press reports that "25% of all Americans are obese, measured by body weight." And Spitzack reports that 79 million Americans are "significantly overweight" (1990, 9). At any given time, 52 million Americans are dieting or contemplating doing so (Spitzack 1990, 9).

Pearlstadt, et al. list a variety of weight reduction therapies. First are the "beauty/health spas and the aerobics, jazzercise, gym and athletic clubs [which] stress exercise and nutritional supplements without much interest in nutrition itself or guidance from health professionals" (1991, 11). The second group is composed of medical clinics and health wellness operations. Their strategy is "nutrition, exercise and psychological support. [They] have clients see dietitians, physicians and nurses" (Pearlstadt et al. 1991, 12). This group is essentially the medical establishment. The third group is the diet centers. They go with the fads, sell special foods and supplements and "have diet plans and strategies while not considering the client's current eating habits, and run maintenance programs while recommending very low calories and rapid weight loss" (Pearlstadt et. al., 1991, 12).

In spite of the variety of programs available and the medical judgment that obesity is harmful to health, do these weight-reduction programs work? Unfortunately, the answer is usually negative. As the *Los Angeles Times* reported in April 1992, "Most Americans who lose weight through commercial programs gain it all back within five years, a blue-ribbon panel to the National Institutes of Health announced . . . saying 20 years of nationwide struggles against obesity have failed." As many as two-thirds of such patients gain lost weight within the first year, the panel said. As Schwartz quotes a physician saying "A person is more likely to recover from most forms of cancer than from obesity" (1986, 227). Spitzack feels that "in the dieting marketplace, lucrative profits override a concern with consumer success" (1990, 14).[1]

Knowledge of this situation has led to a summary "understanding" in the dietetic establishment that diets generally don't work. Such popular weight-reduction books as *Fit for Life* (Diamond and Diamond 1985) advocate permanent dietary change combined with exercise and a healthier lifestyle. Yet the question is still open as to whether there are significantly effective formulas that will help most people lose weight.

From a social/psychological perspective—which has no more hard data support than any of the other obesity therapies—I would like to conceptualize a prospective therapy that might be titled "feminist-praxis." Included in its methodology would be the array of "scientific" techniques of weight reduction and a feminist psychosocial analytic framework.

In this scheme, the first step is the realization that even though one is obese, a sexist society caused it. Change can come from detailing—in the manner of my previous summary of all four of these feminist thinkers—the ways in which society has forced one into an obesity response.

A second element in therapy includes a feminist support group working with others in the same predicament for the same reasons (Orbach's version of the "truth"). Here the love and mutual support—and egalitarian/critical ethos—can provide the substitute for the role that food was serving in one's personality. Instead of mindless reacting, one is actively responding to inequality in a positive, change-oriented (praxis) way. In this context, the anger that plays so great a part in Orbach's analysis of obesity can be redirected toward society. This can take place within the context of the collective anger of the support group.

Another context for creative anger, usually unexplored by the predominantly status-oriented psychological establishment, is social change groups. A variety of women's groups work on freedom of choice for abortion, women's political rights, women's legal rights, equal occupational treatment, desexistizing school texts, and other feminist social-change issues. Though these are not self-consciously therapeutic, they are latently so.

Recall that in addition to the changes in laws that the American civil rights movement created, there were significant psychological changes in, especially, the African American community. Regarded by many as passive, docile, "happy" (like suburban housewives), and "dumb" (the traditional woman's role), African Americans became leaders and workers in sophisticated and personality-enhancing marches, pickets, boycotts, voter registration drives, write-in campaigns, job training and creation, and other change-oriented activities.

Dealing more directly with obesity, self-help/political action/support groups have come together to challenge the pervasive stereotyping and discrimination that occur for fat people. At the same time, by realizing the sexist causes of obesity and having a new degree of acceptance of fatness, praxis groups can perform simultaneous therapy and social action.

In this variant of therapy, one could move to a position where one accepts one's fatness (and perhaps decides to stay fat). Though this is a medically questionable position, it answers the relatively bad success record of reducing programs. Table 5.1 presents an evaluation of the risks of obesity.

Fat pride is analogous to black pride or gay pride. Both justify one's present position and criticize the society that oppresses them. This is the therapeutic mode that the adjustment-oriented medical/psychological establishment usually overlooks. Yet often the most mature personalities come out of these very groups. One would expect a movement for fat women's pride to be a valuable potential tool for psychological liberation.

Table 5.1 RISKY BUSINESS *(Increase in Risk)*			
CANCER	20 to 30% Overweight	40% or More Overweight	Deaths per Year
Male			
Colon/Rectum	26	73	29,100
Prostate	37	29	26,100
Female			
Breast	16	53	39,900
Cervix	51	139	6,800
Endometrium	85	442	2,900
Gallbladder	74	258	5,300
Ovary	0	63	11,600
DIABETES			
Male	156	419	14,859*
Female	234	690	21,927*
HEART DISEASE			
Male	32	95	289,461
Female	39	107	251,857
STROKE			
Male	17	127	61,697
Female	16	52	92,630

*This figure does not include the many diabetics who die of heart disease.

Sources: J. Ckron. Dis. 32:563,1979; Personal Communication, John Lubera, American Cancer Society; Kathy Santini, National Center for Health Statistics.

Reprinted from Nutrition Action Health Letter (Center for Science in the Public Interest, 1875 Connecticut Avenue NW, Suite 300, Washington, D.C. 20009-5728); $20.00 for 10 issues.

Unlike behaviorist theory, the assumption in the psychotherapeutic mode is that once women come to the correct realization (and praxis action) about the reasons for their obesity, they will be in a position where they can actually succeed with the techniques specific to a professionally designed weight-reduction plan. As to a success rate, this strategy has yet to be tried on any significant scale.

An Integrated Model of Eating Disorders

"My boyfriend says, 'don't gain an ounce'" (Debbie, an anoretic woman).

The peculiar position of today's women is to be caught in a schizoid position between fattening food and a thin normative image. On the one hand, the media and friends tell them to be thin is to be beautiful. On the other hand, the media (especially women's magazines) glorify the fattening products of the commercial food industry through articles on dessert recipes and editorial comments on "the way to a man's heart is through his stomach."

Obesity is one response. The other extreme is *anorexia nervosa*—extreme thinness. *Bulimia nervosa* embodies both responses by alternating binging and purging to maintain thinness. In both, culture triumphs over nature (human biology) to the detriment of the person. Bodies adapted 12,000 years ago to store fat consume the junk food of today and "blossom." For some, the only solution to the cultural imposition of extreme thinness is an eating disorder. As Gordon reflects: "Eating disorders are ethnic [cultural] disorders because they magnify a culturally typical solution to a problem that is much more pervasive in the larger culture" (1990, 84).

According to Schwartz, there are at present 1 million anorexics in the United States, 60,000 to 150,000 of whom will die from their disease (1986, 332). "About 5% of high school girls are victims of either anorexia nervosa . . . or bulimia" (Tufts University 1991, 3). And "a full 20% of adolescents are estimated to engage in bulimic behavior" (Tufts 1991, 3). Eating disorders are a major concern on most college campuses.

The American Psychiatric Association classifies anorexics as DSM III-R. They exhibit the following characteristics:

1. A refusal to maintain body weight over the minimum expected for their weight and height.
2. An intense fear of gaining weight or becoming fat
3. A distorted body image of feeling fat
4. Amenorrhea (an absence of menses) (Gordon 1990, 15)

Generally anorexics are between fourteen and eighteen years old. They come from families that made them feel unworthy and in which they were pushed to be high achievers and maintain the appearance of goodness. Gordon cites a similarity with other addictive disorders—in this case a fixation on oral behavior (1990, 18). Because of the open nature of anorexia, Gordon classifies it as a form of rebellion. One need only think of the furor one can cause in a family by refusing to eat. Anorexia has often served as a vehicle for adolescent (usually female) rebellion.

Brumberg summarizes Chernin: "Women who have disordered relationships to food are unconsciously guilty of symbolic matricide and their obsessive dieting is an expression of their desire to reunite or bond with the mother" (Brumberg 1988, 28). The two overriding psychological characteristics of anoretic women are their inability to break their bonds with their mother and an underdeveloped adult sexuality.

Bulimia is defined by the American Psychiatric Association as DSM III. Its characteristics include

1. Recurrent binge eating
2. Frequent purging or severe food restriction between binges
3. Persistent overconcern with body weight and shape (Gordon 1990, 24, quoting the American Psychiatric Association 1987).

Brumberg (1988) writes that bulimics "may experience serious dehydration and electrolyte imbalance" (citing Herzog and Copeland 1985, 296–298; Bemis 1978, 594). When combined with anorexia (sometimes termed bulimarexia) bulimia "can produce hiatal hernia, deterioration of tooth enamel, abrasions on the esophagus, swollen salivary glands, kidney failure, and osteoporosis" (Brumberg 1988, citing Herzog and Copeland 1985; Bemis 594).

Unlike the controlled facade of anorexics, bulimics often come from tumultuous families where alcoholism, drug addiction, sexual promiscuousness, and stealing have been present (Gordon 1990, 21). Bulimics are usually older, 16–20, and enter adulthood with a "false self" (Gordon 1990, 27). Bulimia is five to ten times more prevalent than anorexia and manifests a 9–1 female-to-male ratio. Often the men who are in this category are homosexual. The bulimic's form of rebellion is surface capitulation and "going underground" (Gordon 1990, 13–16). Bulimia is significantly less life-threatening than anorexia.[2] Britain's Princess Diana was reported as being bulimic. Sociologically, the key question is, why the current prevalence of eating disorders?

Gordon cites Devereux (1955) asserting that one can utilize "psychopathology as a way to unravel the mysteries and paradoxes of the culture itself." What do these contemporary illnesses tell us about our culture and society?

Our era is the first in which the image of the thin woman as ideal has been around for more than half a century. Today's young women are the daughters of a generation of weight watchers. They have witnessed their mothers going to weight-loss clinics and attempting to create the kind of unattainable figures that the society idolizes. Thinness, like cleanliness is previous eras, is in fact next to "godliness." As Gordon writes of the addiction to dieting: "The passion, the obsessiveness that surrounds the topic of dieting tells us that we are dealing with an underlying religious idea, but one which masks itself as the epitome of secular rationalism" (1990, 92). Dieting to thinness can replace the Protestant work ethic of previous eras.

Sociologically, eating disorders are on the same continuum as dieting—only taking the practice one step farther. In a society that gives most women contradictory messages, it seems logical that some percentage of women will overdo thinness or evidence the contradictory nature that bulimia so aptly expresses.

Culture, physiology and psychology shape the formulation of eating disorders. Like obesity, eating disorders are subject to a feminist analysis as well. Yet, unlike most obesity, eating disorders, once they get going, are a more serious mental and physical health problem with a physiology and psychology all their own.

They can become addictive. In the same way that cigarette smoking and other drug addictions move from initial unpleasantness to very pleasurable: "In effect, individuals with anorexia nervosa may be dependent on both the psychological and the physiological effects of starvation" (Brumberg 1988, 31).

As Brumberg writes, "As a feminist, I believe that the anorectic desires our sympathy but not necessarily our veneration" (1988, 35). When feminists focus solely on the cultural aspects of these eating disorders, they deny "the biomedical component of this destructive illness by obscuring the helplessness and desperation of those who suffer from it" (Brumberg 1988, 36).

One must further inquire whether these casualties are incidental to a smooth-functioning society. Naomi Wolf doubts it. In her popular *The Beauty Myth*, she explains the psychology and physiology of persistent undernourishment. In many diet centers, women exist on about 1,000 calories daily (Wolf 1991, 195). This is half an average women's normal healthy portion. From such semi-starvation women develop psychological symptoms of "irritability, poor concentration, anxiety, depression, apathy, lability of mood, fatigue, and social isolation" (Wolf 1991, 195). In essence, they experience the same psyche-breaking elements to which slave captains subjected their slaves in their midpassage (Haley 1976).

Similarly, Wolf cites a University of Minnesota six month experiment with thirty-six personality-stable volunteers. Their personalities changed to "depression, hypectirondriasis hysteria, angry outbursts, and, in some cases, psychotic levels of disorganization" (1991, 194). Using a beauty index that is unattainable fosters behavioral characteristics directly contrary to those the women's movement has tried to foster. Thus, dieting becomes a tool for politically destabilizing potentially successful women.

Summary

I have described how body size has varied throughout history. Originally the rich elite were fat. That condition demonstrated, physically, that they ate well. In fact, they ate "organic" produce and meats and little sugar, and they exercised in their normal day. They just ate tremendous amounts. Today we are subject to the reign of the ultra-slim body style that began with "flappers" in the 1920s, and has continued with minor variations into the present. It is supported by government policy and private advertising.

The presence of AIDS has raised some questions about ultra-slimness. One sees an anorectic and thinks of someone dying of AIDS. As a result, a rational society would set a standard that is closer to biological normality.

In 1983, life insurance companies began to accept a little heavier weight-to-height ratios. The University of California *Berkeley Wellness Letter* reports that in November 1989 the U.S. Food and Drug Administration and the Health and Human Services Department "revamped the federal guidelines for body weight." The new reference point is "healthy weight," rather than a cosmetically desirable weight. The new guidelines are shown in Table5.2.

Table 5.2. Suggested Weight for Women (without clothes or shoes)

Height	19-34 Years	35 Years and over
5'0"	97–128	108–138
5'1"	101–132	111–143
5'2"	104–137	115–148
5'3"	107–141	119–152
5'4"	111–146	122–157
5'5"	114–150	126–162
5'6"	118–155	130–167
5'7"	121–160	134–172
5'8"	125–164	138–178
5'9"	129–169	142–183
5'10"	132–174	146–188
5'11"	136–179	151–194
6'0"	140–184	155–199
6'1"	144–189	159–205
6'2"	148–195	164–210
6'3"	152–200	168–216
6'4"	156–205	173–222

Source: University of California, *Berkeley Wellness Letter*, February 1991:2.

Were this information widely disseminated, women might have a better opinion of themselves. But in a society where males can profit from creating dissatisfaction among women, one cannot really expect that this scientific knowledge will be communicated with the rigor comparable to that with which the media parade slim bodies. As long as a woman's *appearance* is more important than her being, most women will be influenced in large degree by the cultural image of the anoretic model. One can only hope that schools will provide a counterbalance to the commercial media. As the *Berkeley Wellness Letter* states: "Great bodies come in many shapes" (2/91:1).

An examination of world images of beauty and eating lead to further questions about the beliefs of various people. We therefore move to a discussion of culture and food.

References

Bailey, Carol A. 1990. "The Role of Dieting in the Etiology of Bulimia." Abstract of paper presented at the fourth annual meeting of the Association for the Study of Food and Society. Philadelphia, Pennsylvania, June.

Bell, Rudolph M. 1987. *Holy Anorexia.* Chicago: University of Chicago Press.

Beller, Ann Scott. 1977. *Fat and Thin: A Natural History of Obesity.* New York: Farrar, Straw & Giroux.

Bemis, K. 1978. "Current Approaches to the Etiology and Treatment of Anorexia Nervosa." *Psychological Bulletin 65*: 593–617.

Bennett, William I. 1984. "Dieting: Ideology Versus Physiology." *Psychiatric Clinics of North America.* 7(2):321–334.

Boodman, Sandra. 1992. "Researchers Link Hormone to Bulimia." *Grand Rapids Press*, May10, B8.

Brownell, Kelly D. 1982. "Obesity." *Journal of Consulting and Clinical Psychology 50*:820.

———. 1984. "The Psychology and Physiology of Obesity: Implications for Screening and Treatment." *Journal of the American Dietetic Association 84*(4):406–413.

Bruch, Hilda. 1979. *The Golden Cage.* New York: Vintage Books.

Brumberg, Jane Jacob. 1988. *Fasting Girls: The Emergence of Anorexia Nervosa as a Modern Disease.* Cambridge, Mass.: Harvard University Press.

Chernin, Kim. 1981a. *Womansize: The Tyranny of Slenderness.* London: Woman's Press.

———. 1981b. *The Obsession: Reflections on the Tyranny of Slenderness.* New York: Harper & Row.

———. 1985. *The Hungry Self: Women, Identity, and Eating.* New York: Harper & Row.

"Congressman Wants Controls on Diet Plans." 1990. *Grand Rapids Press*, March 26, A10.

Daufin, E.K. 1991. "Fat People Are the New Targets of Open Prejudice." *Grand Rapids Press*, March 26, A8.
Delatt, Ken. 1985. "Eating Disorders." Class talk at Aquinas College, Grand Rapids, Michigan.
Devereux, George. 1955. "Normal and Abnormal." In *Basic Problems of Ethnopsychiatry*, ed, George Devereux. Chicago: University of Chicago Press.
Diamond, Harvey, and Marilyn Diamond. 1985. *Fit for Life*. New York: Warner Books.
"Does Weight Harm Your Health?" 1987. *Nutrition Action Health Letter*, January/February, 7.
"Experts Call for Diet Program Reforms." 1992. *Grand Rapids Press*, October 30, D9.
Fiddes, Nick. 1991. *Meat: A Natural Symbol*. London: Routledge.
Fieldhouse, Paul. 1986. *Food and Nutrition: Customs and Culture*. New York: Croom Helm.
Gordon, Richard A. 1990. *Anorexia and Bulimia: Anatomy of a Social Epidemic*. Cambridge, Mass.: Basil Blackwell.
"Great Bodies in Many Shapes and Sizes." 1991. University of California, *Berkeley Wellness Letter*, February, 1–2.
Haley, Alex. 1976. *Roots*. Garden City, N.Y.: Doubleday.
Harris, Robert T. 1983. "Bulimarexia and Related Disorders." *Annals of Internal Medicine 99*(6):800–807.
Herzog, D. 1982. "Bulimia: The Secretive Syndrome." *Psychosomatics* 27:481–487.
Herzog, David B., and Paul M. Copeland. 1985. "Eating Disorders." *New England Journal of Medicine 313*(5):295–303.
"How Fat We Are." 1990. *Detroit Free Press*, April 1, 6.
Levin, Jack. 1988. "Fat Chance in a Thin World." *Bostonia*, January/February, 61–62.
Meadow, Roslyn M., and Lillian Weiss. 1992. *Women's Conflicts about Eating and Sexuality*. New York: Harrington Park Press.
Millman, Marcia. 1980. "When I'm Thin I'm Perfect." *Savvy*, 32–36.
"Most Who Lost Weight on Diet Programs Gain It All Back, Panel Says." 1992. *Grand Rapids Press*, April 1, A7.
"Obesity Figures High in U.S." 1992. *Grand Rapids Press*, March 29, B7.
Orbach, Susie. 1978. *Fat Is a Feminist Issue*. New York: Berkeley.
Pearlstad, Harry, Laura Kem, Judith Anderson, and Karen Petersmark. 1991. "Differences among Commercial Weight Loss Clinics." Paper presented at the meeting of the Association for the Study of Food and Society, Tucson, Arizona, June 14.
"Real People Eating Real Food." 1986. *Grand Rapids Press*, May 5, C5.

Ritenbaugh, Cheryl. 1982. "Obesity as a Culture-Bound Syndrome." *Culture, Medicine, Psychiatry. 6*:347–361.

Robbins, Jotin. 1987. *Diet for a New America*. Walpole, N.H.: Stillpoint.

Robbins, Carole J., and Steven J.C. Gaulin. 1990. "Fat and Thin in Evolutionary Perspective." Paper presented at the meetings of the Association for the Study of Food and Society. Philadelphia, June.

Rozin, Paul, and April Fallon. 1988. "Body Image, Attitudes to Weight, and Misperception to figure Preference of the Opposite Sex: A Comparison of Men and Women in Two Generations." *Journal of Abnormal Psychology 97*(3):342–345.

Schwartz, Hillel. 1986. *Never Satisfied: A Cultural History of Diets, Fantasies, and Fat*. New York: Free Press.

Sieger, Jim, producer. 1985. *The Wasit Land*. Gannett Corporation, Inc. & MIT Teleprograms, Inc. Film.

Sobal, Jeffrey. 1984. "Group Dieting, the Stigma of Obesity and Overweight Adolescents: Contributions of Natalie Allon to the Sociology of Obesity." In *Obesity and the Family*, ed. David J. Kalen and Marvin B. Sussman. New York: Hayworth Press.

———. 1984. "Marriage, Obesity, and Dieting." In *Obesity and the Family*, ed. David J. Kalen and Marvin B. Sussman. New York: Hayworth Press.

———. 1987. "Socio-economic Status and Obesity: Application of a Social Selection and Social Causation Theories." Paper presented at the meeting of the Association for the Study of Food and Society, Grand Rapids, Michigan.

———. 1988. "Public Beliefs about Fattening and Dieting Foods." Paper presented at the 1988 meeting of the Association for the Study of Food and Society. Washington, D.C., June.

———. 1991. "Obesity and Socioeconomic Status: A Framework for Examining Relationships between Physical and Social Variables." *Medical Anthropology 13*(3).

———. n.d. "Obesity and Nutritional Sociology: A Model for Coping with the Stigma of Obesity." Paper for New York State Hatch Project NY(C) 399404.

Sobal, Jeffrey, and Albert J. Stankard. 1989. "Socioeconomic Status and Obesity: A Review of the Literature." *Psychological Bulletin 105*(2):260–275.

Sobal, Jeffrey, Carmine M. Volentie, Herbert L. Muchie, Jr., David Levine, and Bruce R. DeForge. 1985. "Physicians' Beliefs about the Importance of 25 Health Promoting Behaviors." *American Journal of Public Health 75*(12):1427–1428.

Spitzack, Carole. 1990. *Confessing Excess*. Albany: SUNY Press.

Squire, Susan. 1983. "Is the Binge-Purge Cycle Catching?" 1983. *MS Magazine*, October, 28–31.

Suplee, Curt. 1991. "Moms Linked to Daughters' Eating Habits." *Grand Rapids Press*, May, F3.

Tobias, Alice. 1980. "Social Consequences of Obesity." *Journal of the American Dietetics Association* 76:338–343.

Van Lenten, Jay. 1986. "Eating Disorders." Talk at Aquinas College, Grand Rapids, Michigan, Spring.

Van'T Hof, Sonja E. 1991. "Never Proven but Yet Explained: The Increase in Anorexia Nervosa." Paper presented at the American Sociological Association, Cincinnati, Ohio, April 25.

"When Growing Pains Hurt Too Much: Teens at Risk." 1991. *Tufts University Diet and Nutrition Letter*, June, 3–6.

Wolf, Naomi. 1991. *The Beauty Myth*. New York: Morrow.

Wooley, Susan, and O. Wayne Wooley. 1986. "Thinness Mania." *American Health*, October, 68–74.

Notes

[1] 1.Recently, state and national governments have begun to take some responsibility for the freewheeling profits of the diet and reducing industry. "In . . . 1990, a U.S. House Subcommittee met to hear discussion of 'Deception and Fraud in the Diet Industry'" (Pearlstadt et al. 1991, 1). This resulted in the U.S. Department of Health and Human Services and the National Institute of Health Nutrition Coordination Committee issuing a report suggesting that the National Institute of Health judge the effectiveness of weight loss clinics.

In Michigan, a task force of the Michigan Health Council published in 1990 *Toward Safe Weight Loss: Recommendations for Adult Weight Loss Programs in Michigan* (Pearlstadt et al. 1991, 3).

It remains to be seen how much government will attempt to regulate the profits of this ubiquitous industry.

[2] 2.There is some recent evidence of possible hormonal involvement with bulimic behavior. Scientists at the National Institute of Mental Health have noted that bulimia "appears to be associated with excessive levels of the brain hormone vasopressin."

Chapter 6

Culture and Food

American Tourist: I'm appalled! You Indians have all these lazy, flea-bitten cows running around the streets of New Delhi and you have hundreds of starving beggars. Why don't you eat these cows?

Hindu Guide: We value the cow as holy. No one eats cows. It would be sacrilegious.

American Tourist: But you have so many starving and undernourished people. You could grind up these old cows and make them into hamburgers which you could sell cheaply to the poor. You would save many of them from dying of hunger.

Hindu Guide: But that would violate the highest principles of our religion. Better to keep the religion. If these beggars live a righteous life, they will be reincarnated in a better caste. Death for them is just a release from their suffering. They prefer it this way.

What is at issue in this dialogue? Behind the interchange are social and cultural analyses that have long traditions in anthropology, history, folklore, and (much shorter) sociology.

How does one explain this dialogue? The contrasting answers contain the essence of *idealist* (cultural) and *materialist* (societal) analysis. And these modes of analysis provide the theme for this chapter as it examines not only the Hindu sacred cow but also Jewish food proscriptions, Thanksgiving dinners, Indian and Chinese cuisine, and fast food. In each subject area, I provide both an idealist and a materialist analysis, usually favoring the latter.

The Indian Sacred Cow

In analyzing the opening dialogue, one can find the "ugly American" tourist offering a practical solution to an "irrational" situation. From the tourist's ethnocentric perspective (from a culture where meat is its stable and status food), the Indian system is irrational. If cows are running around in the midst of mass human starvation, would it not make more sense to eat the cows?

The Hindu guide's answer is from an *idealist* perspective. He maintains the *autonomy* of the religious idea of the holiness of cows and the Hindu (as well as Buddhist, Jain, and Sikh) proscriptions about the consumption of meat. His general point is that the religious idea (the basis of idealism) needs no other explanation than that it is the word of the holy beings. The situation is analogous to Israelites not eating pork solely because the Old Testament commands it.

This mode of argument is usual for religious zealots and for conservatives in general. It assumes that ideas have a life of their own. But it is fundamentally antisociological.

Sociology is social *science*. As such, it seeks to reveal what is behind the appearances (values) in society. To do this in the case of Hindu prohibitions on meat consumption involves a functional material analysis of the role of the cow in Indian society.

First, the dung from the cattle is valuable in India. India's poor have pretty much denuded the countryside in the search for firewood for cooking and heating. The poor can burn dung as fuel for heat and cooking. In addition, dung is fertilizer. In a country where the poor have no capacity to purchase petroleum-based fertilizer, dung provides a source that can be freely collected and used to grow grains, fruits, and vegetables. A third function of the Indian cow is plowing power. Cows (and oxen) can be hitched to plows in the traditional agrarian manner. In fact, animate energy sources have been the mark of agriculture since its beginning. It is only in the past century that the developed world has utilized petroleum fuel in internal combustion engines to plow fields. Fourthly, because cows can feed on plants that humans do not eat (usually known as "weeds"), Indian cows need no grain resources. They eat the unfenced weeds that grow up around houses and help keep the grass down in the public parks. They are also fed on the otherwise useless by-products of agriculture. Fifth, in addition to weeds, cows scavenge refuse. This reduces the organic material that lies around and must be disposed of. Sixth, cows keep down the speed of traffic in cities like New Delhi. Because so many cows roam the city, it is next to impossible to maintain high speeds of auto or scooter driving. Because hitting cows damages these vehicles, one must exercise prudence in driving. Seventh, by prohibiting the slaughter of cattle, therefore allowing a surplus of them, the price of cows is kept low. This makes it possible for a peasant or poor person who needs a cow to purchase it at an exceptionally low price. Eighth is the food value. Although Hindus, Jains, and Buddhists are vegetarians, they are lacto-vegetarians. They consume milk products. Yogurt and milk, coming from a holy animal, are highly valued in the Indian cuisine. Traditionally a *raita* (yogurt, tomato, cucumber, and/or raisins with Indian spices) accompanies most lunch and dinner meals, serving the function of a salad. And a yogurt drink, *lassi*, is used for refreshment.

Traditionally, *ghee* is made of "clarified butter." One heats spices in *ghee* and mixes them with *dahl*. Also *ghee* is the medium in which vegetables are stir-fried. It is the generic oil of Indian cooking.

Ninth is the cow's function as a means of transport and a beast of burden. Although not ideal for this function, cows can be ridden. They can also be loaded with goods for movement from one place to another. Tenth, cows secrete methane gas. At present, the world's 1.3 billion cattle contribute 70 million tons of methane gas each year to the atmosphere. This is one-fifth of the greenhouse gasses. This is an energy source waiting for correct methods of harvesting.[1] Eleventh, when cattle die of natural causes, the lower castes in India eat the meat. In addition, the skin is used for leather in the same manner as in the Western world. But in India, one must wait for a cow's natural death. The final disposal is simply postponed. The twelfth function of not eating the meat of the cattle is spiritual. Generally it has been the vegetarians and fasters of the world who have reported the most spiritual experiences. Even in the Western world, it is well known that Jesus fasted for forty days.

Similarly, Goody reports "a firm belief . . . that a man is what he eats and that purity of thought depends on purity of food" (1982, 115). In the Indian caste system, "to move upwards meant changing one's diet, usually by becoming more vegetarian" (1982, 115). It is assumed that one who eats in a vegetarian manner or one who fasts is more spiritual than one who does not.

In summary, this materialist-functionalist analysis demonstrates the overwhelming practicality of the Indian relegation of cattle to the realm of the "sacred." The spiritual practice has a multitude of practical material justifications.

Though our Hindu guide defends the practice of not killing cattle on religious grounds, these grounds themselves derive from a material base. As K.N. Nair (1987) writes:

> The avoidance of cattle slaughter originated in the Vedic period (1200–900 B.C.). The strategy adopted was to restrict animal sacrifice to ceremonial occasions and the entertainment of guests and to make the cow a sacred animal (Harris 1980; Srinivasan 1979). The sacredness of the cow was ensured by emphasizing the importance of milk in the diet, the cow's divine origin, the belief in the transmigration of the soul and the cow's role in helping humans along the path to salvation (in Harris and Ross 1987, 449).

The dominant religious idea (cited by idealist social scientists) is, in fact, grounded in the material needs of the society. Culture, as a system of ideas, beliefs, and values, began as an adaptive response to the material situations which people encountered in their societies. (See Box 6.1.) This materialist analysis demonstrates the sociological basis of religious beliefs. Though our Indian guide may be a practicing Hindu, he is not a sociologist.

Box 6-1 Sociological Methodology

Sociologists and anthropologists conceive *culture as* the ideas, beliefs and values which constitute the "superstructure" (a marxian term) of society. Culture is the sum of the non-material factors in society.

Culture can be seen from a macro and a micro perspective. Macroscopically, culture is equivalent to the German *weltanschauung*. It is a total world view. It is how a (bounded) society sees and interprets the outside world.

Microscopically, culture is composed of specific ideas, beliefs, and values as they are expressed in the norms, mores, folkways, and laws of a society. Thus, the idea of freedom of speech is valued by our society and, within limits, protected by the American constitution and the culture mores of the society. Such an institution as the "Letters to the Editor" section of many newspapers embodies the idea that the average person (who writes in a way which the editor considers responsible) should have a chance to freely communicate her/his opinions in the privately owned mass media of the newspaper.

Traditionally, anthropologists have focused more on the study of "culture," while sociologists have come from the perspective of "society." Nevertheless, all societies have value systems and most value systems have some societal embodiment.

Contemporary sociological texts usually mention the terms "ethnocentrism" and "cultural relativity." The former is an ethical situation where one views a "foreign culture" from one's own value system. The latter is a value-neutral position where one simply accepts the practices of other cultures.

It has been fashionable in American sociology to adopt a value neutral position. While this may avoid the gross intercultural misunderstandings (ethnocentrism) that our American tourist (in the dialogue) espouses, it often leaves one in a position of valuing hunger and death because they seem to be accepted in some parts of the world.

Our analysis will raise some questions about the seemingly "value neutral" stance. This question becomes especially important in discussions about world hunger which are raised in chapter 9 of this book. In an attempt at analytic clarity, I have focused chapter 6 on "culture" and left "society" for chapter 7. Nevertheless, they are indivisible. That we focus on culture here should not be to the exclusion of society. In fact, I argue in this chapter that culture is more dependent on the material factors of society than vice versa.

Until quite recently, food has been studied in the domain of history. Tannahill's *Food in History* (1973) and Norbert Elias's *The History of Manners* (1940) are two major works that deal in an historical and comparative manner with developments in food, culture and society. Nevertheless, there is a tradition in anthropology of materialist analysis. In food, its most able proponents are Marvin Harris (*Good to Eat*, 1985) and Sydney W. Mintz (*Sweetness and Power*, 1985). These authors form my personal intellectual heritage.

Jewish Pork Proscriptions

If a materialist–sociological analysis is the key to understanding eastern Hindu scriptures, how might it be applied to Western religious proscriptions? For instance, what is the explanation of why Jews are forbidden to eat pork?

Once again, we start with the obvious explanation from the religious texts. Specifically, as stated in the Old Testament book of Leviticus (and still practiced by orthodox Jews), the people of Israel are forbidden to eat animals that do not ruminate (chew the cud) or do not have cloven hooves. Pigs (swine) have cloven hooves, but they do not ruminate.

For orthodox Jews, the question is simple. They must obey the written word of God. Therefore, they follow this (among other) proscriptions. The word of God is what counts. This is the first *idealist* interpretation of Jewish avoidance of pork. It has guided the Jewish community down to the present. A second idealist interpretation is provided by the English structuralist anthropologist, Mary Douglas. Summarily, after dismissing a number of materialist attempts, Douglas elaborates the logical consistency involved. She finally justifies the whole scheme in a manner similar to Jews themselves. That is, Jews must do what is holy. And what is holy is a neat, consistent classification.

In opposition to the idealist and static analyses of the avoidance of pork eating, Marvin Harris (1985) and Ann Murcott (1986) provide sociological accounts. Harris highlights nutrition and the notion of an ecological niche. Murcott focuses on group identity.

The most popular materialist analysis by lay people of the Jewish pork proscriptions comes from our contemporary knowledge of trichinosis. If pork is not thoroughly cooked, one can get trichinosis from it. Like so many other foods in the history of humankind, the people who ate correctly survived. About those who did not, we know very little (because presumably, they died from faulty eating habits). Ancient societies did not need to have a medical knowledge of trichinosis to notice that people got sick from uncooked pork (Harris 1985, 70).

More important is the ecological niche which pigs did and did not fill in Middle Eastern society. Specifically, of the animals of the Middle East, "cattle, sheep and goats are ruminants, the kind of herbivores which thrive best on diets consisting of plants that have a high cellulose content... [such as]... grasses and straw" (Harris 1985, 72). This meant that these animals did not compete with humans for grain or meat. In addition, many produced milk; sheep produced wool. All provided leather, and dung for fertilizer and some pulled plows. And they did not require artificial shade or extra water in which to wallow (to cool off) as pigs do.

Furthermore, Harris notes that the book of Leviticus was written late (450 B.C.) in Israelite history (1985, 79). Because the previously verdant Middle East had by that time been denuded, the pig was inappropriate to environmental conditions of lack of shade and high cellulose grazing. Harris writes that "the food laws in Leviticus were mostly codification of pre-existing traditional food prejudices and avoidances" (1985, 79).[2]

Another function that Jewish proscriptions about pork have (along with other Jewish food practices) comes from social psychology. Ann Murcott feels that, regardless of the specific theological justifications for the Jewish food practices, their ultimate function is that of providing social solidarity. Specifically, "attention to a people's classification of the natural world and to the place human beings have in it provides an explanatory 'template' that makes sense of habits which otherwise, from our own point of view, appear to have no sense. . . . The daily activity of eating thus embodies answers to the question of Israelite cultural identity" (1986, 21).

Additionally, the Israelites sought to differentiate themselves from certain Babylonian groups because these groups worshiped the pig god, Ishtar. Israelites felt that eating pork would be an identification with Babylonian religion (Lewis 1990).

In summary, the idealist explanations seem rationalizations for practices developed in eminently practical and rational material situations. As a social scientist, one need not take religion as an exclusive determiner. Scientifically discoverable material factors caused or influenced the development of this religious doctrine.[3]

World Cuisines: China and India

Internationally, probably the two most popular cuisines are those of China and India. I first summarize their similarities and then note some of their peculiarities.

Nutritionally, both societies ate mostly vegetables and grains. Neither had the high meat, high cholesterol bias that American and Western food generally manifests. And their older citizens do not die as regularly from the circulatory diseases of heart attacks and strokes.

Both Indian and Chinese cuisines make artistic use of spices. But spices serve two material functions: making a bland and repetitious diet interesting and disguising the taste of partially rotted food (from the days before food preservation by canning and cooling). The Chinese and Indians spice differently. The ability to mix spices is the mark of a good cook.

Both the Chinese and Indian spice legumes (soybeans in China; peas and lentils in India) combined with rice and some wheat in their basic diet. Although Chinese and Indian spices differ, they serve these two simultaneous functions. Such an acceptable common food as chutney in India is, in fact, a spiced, partially rotted fruit. In addition, in both countries, food is conceived as medicinal. As Julie Sahni writes:

> The role of spices and herbs goes far beyond pleasing the palate and soothing the senses. They have medicinal properties that were known to the ancient Indians. Ayurvedic scripts in the three-thousand-year-old Holy Hindu Scriptures on herbal medicine list the preventive and curative powers of

> various spices, herbs and roots in the treatment of common physical ailments. . . .
>
> The Holy Hindu Scriptures also document the effect of spices on body temperature. Spices which generate internal body heart are called "warm," and those which take heat away from one's system are called "cool" spices. Bay leaf, black cardamom, cinnamon, ginger powder, mace, nutmeg, and red pepper are "warm" spices and are recommended for cold weather.
>
> Spices also induce perspiration which helps one to feel cool and comfortable. This is why Indians prefer to drink piping hot spiced-laced tea in hot weather (1980, 3–4).

The sophisticated Indian cook must know the interactions of all the foods and spices in any given dish.

In China, "all foods are considered to have curative value" (Lai 1978, 11). For example, "the Cantonese are particularly prone to treat food as medicine sometimes as 'hsueh' (damaging); some as 'liang' (cool) and some as 'je' (hot)" (Lai 1978, 11).

This demarcation of hot and cool foods in Indian and Chinese cuisine/medicine is also to be found in much of Latin American culture. Western science is often inclined to dismiss these schema as "prescientific." Yet scientific research into the curative powers of Chinese herbs—as well as those of Native American Indians—have yielded some effective medical results. Perhaps it is our ethnocentric Western "scientific attitude" that inhibits our learning from civilizations that predate those of the West. According to Anderson, "the whole concept of a medical therapy based on gentle, inexpensive, everyday means of strengthening the body and soothing its aches has much to contribute to our modern system with its powerful and dangerous remedies that all too often create iatrogenic pathologies of their own" (1988, 243).

As McIntosh writes, "Harris notes that Western science has found, upon examination, many 'irrational' food practices have medicinal value."

Additional similarities between Chinese and Indian cuisines include cooking in a wok and a disinclination to use leavened bread. The latter may come from times when yeast would not keep. The former is evidence of a material situation in which fuel is scarce. By cutting food into smaller parcels, the cook could heat it much faster in a wok. This method uses much less fuel.

Each nationality has its own peculiarities. We have already reviewed the religious and material basis for the Indian aversion to beef (as well as that of Jews for pork). And we have noted that the role of complete protein in the Indian diet is through dairy products: yogurt and *ghee* (clarified butter).

From a materialist/evolutionary perspective, one must assume that Indian people who were lactose tolerant (not a majority condition in the world) were selected to evolve in a society where meat was not much eaten. Milk products are an excellent source of complete protein in themselves. They also, when eaten with an incomplete protein, complete it. Unlike the Indian population, most Chinese are lactose intolerant. They do not produce the enzyme lactase, which can digest lactose. How do they get their calcium? The answer is the rest of the Chinese diet. Calcium is also to be found in dark green, leafy vegetables and in soybeans. Both are a large part of the Chinese diet. And the Chinese get lots of vitamin D from the sun. Vitamin D stimulates the absorption of calcium.

Another major difference that characterizes Chinese eating is pork consumption. Because there is no need to milk animals (pigs are unmilkable), pork could only be a source of meat. In addition, because pigs are more efficient converters of cellulose, they can reuse (eat) human feces, as can dogs, which the Chinese also eat (Anderson 1988, 125). In addition, housing pigs adjacent to the house made them ideal consumers of household garbage. This includes the stems from the green leafy vegetables. And because Chinese agriculture is more human labor intensive (involving a lot of terracing and irrigation), there was not a large need for the draft function of the cow or ox.

Furthermore, partly as a function of the need for many agricultural laborers, Chinese dinners were affairs of large families. Traditions of eating were of the communal meal around a large table with great varieties of foods. Everyone would like some of it. And group dining reinforced group solidarity.

Finally, one notes the Chinese custom of eating with chopsticks. While Europeans were still eating with a knife (a "mini-sword") and their hands, and while Indians still often eat with their hands, the Chinese saw themselves as the elite of eaters. As Lai writes: "The Chinese invented chopsticks and thus transcended the animal realm. Nothing is more symbolic of the Chinese intellect than chopsticks. They represent the unity of duality, the practical application of Yin and Yang, the pristine demonstration of human skill from which all latter-day technology sprang" (1978, 76).

One can analyze Chinese and Indian cuisine from the perspective of idealism. By looking at their distinctive tastes and cooking methods, one can admire the uniqueness and idiosyncrasies of Asian "culture." Most of us do that when we go to a Chinese or Indian restaurant to partake of their respective cuisines.

As a social scientist, however, one must note that a materialist analysis remains the best explanation for most cuisines. Ecological conditions (lack of fuel), efficient use of waste (the pig), and the multitude of reasons for not slaughtering the Indian cow form the material basis for the development and the continuance of these practices. Nevertheless, these material bases should not detract from our admiration of the creative, volitional input of Asian cooks and herbalists. Within the broad outlines of material necessity, there is indeed room for creativity in cooking and in creating the unique and wonderful tastes that make Asian cuisine such high art compared to most Western food practices. As Camp writes: "What is central, and perhaps irreducible, is the humanizing aspect of the cookery process, which alters the perception of the foodstuff in question from a thing of the natural order to a thing of the human order" (1989, 95).

High cuisine is a tribute to human creativity and cultural inventiveness. The search for art and variety in food is probably why, in the era of internationalization, Chinese and Indian restaurants have been the first and most pervasive "foreign" restaurants to open. Their level of cuisine, combined with their ability to provide healthy and reasonably priced meals from basic, mostly inexpensive, ingredients, make them one of the best cultural and nutritional bargains. And in an increasingly international world, "by ingesting the foods of each new group, we symbolize the acceptance of each group and its culture" (Kalcik 1984, 61).

American Food I: Thanksgiving Dinner

From the perspective of our previous explanations of the superiority of materialist analysis, Thanksgiving is probably the least usefully analyzed as a material necessity. The only materialist "debunking" function that is really useful is to note that in opposition to the ideological justifications that currently underlie it, Thanksgiving was in the New World a partial repetition of the ancient English Harvest Home Festival. Instead of a religious holiday, the Pilgrims opted for "a time of joy, celebration and carousing, far removed from any suggestions of solemn religious concern" (Deetz 1972, 30).

In English history, Harvest Home was the most important (and the most rowdy) of rural festivals and was a celebration of a successful harvest (Deetz 1972, 35). Christians continue to celebrate this origin in a Thanksgiving hymn that opens with the lines "Come, ye thankful people come, raise the song of harvest home." And the hymn closes both first and last verses with "Raise the song of harvesthome." The hymn was written by Herry Alford and George J. Eluey around 1840. The nineteenth-century church seems to have wanted historical continuity.

But, though its historical origin was the Home Harvest Festival, Thanksgiving has evolved into the best of American holidays. It lacks the commercialism of Christmas and the hypocrisy of Easter. It is a tribute to the artistry and strength of American cultural (legitimately analyzed as idealist) tradition.

From a nutritional point of view, Thanksgiving now seems an overabundant holiday that, more than any other, focuses on eating all the traditional foods. It may be compared to the *potlatches* which characterized many tribal cultures (see Box 6.2).[4]

Box 6.2. Potlatches

Traditionally the *potlatch* was thrown to celebrate the affluence of the giver. By the standards of the little society in which it was performed, food and material goods were abundantly redistributed. The higher the social status or the richer the one performing it, the more food was to be displayed, distributed, or wasted.

In many of these societies, this rather massive passing out of free food provided a goodly amount of protein for people who might not get enough on a regular basis. In addition, its social function might involve the obligation of exchange. If one (of relatively high status) attended another's *potlatch*, then at some time in the future, one might be expected to have one of one's own and to invite (reciprocally) the person who had performed the previous one.

Thanksgiving provides a rich field for idealist cultural and societal analysis. In the categories of folklorists, it is a "food event" (Camp 1989, 56) or "foodway" (Yoder 1972, 32). That is, it is a multifaceted food occurrence that can be culturally and socially deciphered. Regardless of its actual historical origin, it has come to be syncretistically tied to the religious giving of thanks.

Thanksgiving has a specific form of food presentation. Correlated with the performance of Thanksgiving are certain social functions that are useful to elaborate. Charles Camp sets the project of the analyst of foodways as "to make explicit about food that which is implicit in American culture" (1989, 14). Specifically, Camp locates his theoretical base in William Graham Sumner's "folkways." He defines them as "customs, practices, ways of thinking shared by members of the same group" (1989, 24). Therefore, foodways are "the intersection of food and culture: all aspects of food which are culture-based, as well as aspects of culture which use or refer to food" (1989, 24).

Like a birthday celebration, most American Thanksgivings have a certain identifiable coherence. Students in my Food and Society course reflected on their Thanksgiving experiences. Diane De Boer noted that it "brought back a lot of memories of Thanksgiving pasts" (1990, 1). And Jessica Wuroky wrote about her first visit to her new husband's family: "Strangely enough I felt comfortable. I guess different families are not so different when sharing an old familiar custom such as Thanksgiving dinner" (1990, 1).

Specifically, most students noted traditional Thanksgiving foods: turkey, cranberry, stuffing, mashed potatoes, sweet potatoes, gravy, olives, salad, and pumpkin (and often mince) pie. Ritually, there was a prayer (with an often guilty mention of those who were "less fortunate") and a social situation of being on one's best behavior. De Boer wrote: "In conversation around the table everyone is aware of how many poor people there are that desperately need our help. Whenever the topic was brought up we realized how lucky we are and the subject was changed right away. I'm sure this was because of the guilty feelings" (1990, 2).

A relatively traditional part of Thanksgiving has been sport and revelry, which were part of the English Harvest Home Festival and have become a part of the American version. One student mentioned her family's going out to play basketball. Freund cites flag football (1991, 313). But more common, I suspect, is the tradition of watching football on television. In some "sportaholic" homes, the game schedule dictates the timing of the meal.

Of equal import is the tradition of sharing the meal: "Commensality—the sharing of food—establishes communion and connection in all cultures" (Counihan 1989, 359). And Camp writes that food events are the "intertwining of human business—the serving of food as a customary gracenote, the expression of holiday hospitality, bread broken together symbolizing shared feelings" (1989, 57). The Thanksgiving meal is the actual performance of group unity.

Student Karen Dean observed the social/cultural value of sharing:

> Adults sampled everything, unless they couldn't eat something due to allergies or extreme dislike. Adults piled food upon their plates, which differed from the children's plates with little food and empty spaces. Many adults ate everything that they took even though they may have not liked something. The feeling of fullness seemed to be a common goal of the adults and seemed to signify satisfaction and approval of the dinner. No matter how "stuffed" though, dessert was always readily accepted. (1990:4)

In addition, in the context of the rapid dissolution of American family dinners (or any meal), Thanksgiving also recalls traditional family experience. Student Roxanne Vaz commented on her first Thanksgiving, "It was nice to see the entire family, immediate and extended, sit together and enjoy a meal. This was what I was used to back in Sri Lanka, a family meal. Since I arrived here I never had one" (1990, 1). And student Robin Sapp noted, "Thanksgiving was the only time and first time that I was able to spend time with both my mother and father's families since they were divorced" (1990, 1).

The Thanksgiving meal provides a socialization experience in cooking. In America, the passing on of cooking practices between mother and daughter is summed up in the multitude of food that must be cooked for a Thanksgiving feast. Possible culinary embroidery on the basic Thanksgiving pattern would include the making of cranberry jelly (from scratch) and family preferences for the contents of the turkey stuffing. Like medieval craftspeople, mothers pass on the craft and art of making these Thanksgiving dishes to their children during the meal preparation. It also seems that this is one of the few remaining experiences for men and women to learn and practice the "setting of a table."

Thanksgiving provides an opportunity for cultural expressions of family, social, and national (religious) solidarity. It unites small and large groups and expresses their values. It is a "food event" par excellence, which bears continued study as do Christmas, Easter, birthdays, anniversaries, weddings, funerals, festivals, and state dinners.

American Food II: Fast Food

If Thanksgiving, with its cooking, artistic traditions and homemade pumpkin pie, is some of the best that American culinary culture has to offer, fast food is the opposite. Turkey dinners are part of the individualized, craft tradition of Western culture. Fast food is the ultimate "rationalization in the interest of profit" (Hell 1989). The sociological reader should note the direct line of analysis to the historical sociologist Max Weber.

Fast food connects back to the theme of oligopoly, highlighted in chapter 1. Until quite recently, fast food restaurants fried most of their offerings in the highest cholesterol fats (lard or palm oil) because it was cheaper and lasted longer. Only as the result of negative advertising have they switched to supposedly healthier oils. Unfortunately, current research is finding that the trans fatty acids in the current cooking oils "raise cholesterol, possibly as much as saturated fats do" (Wooten and Liebman 1993, 10).

The key to successful operation among the oligopolists of fast food is cheapness, industrial efficiency and advertising. The biggest, McDonald's, is described by one owner as a franchise operation in which the person with the franchise contributes 5 percent of his receipts to McDonald's in return for national advertising and a McDonald's franchise (Heller 1989).

Although, in one sense, fast food fulfills the socialist ideal of the fast, efficient, low-cost communal kitchen, in its present form it undermines its customers' nutritional status. By specializing in sugar, salt, frying (even low cholesterol fat adds calories), and meat, fast food makes a travesty of the socialist ideal of having healthy food available to the masses. Only Wendy's and Rax offer viable salad bars. Some Wendy's franchises offer Mexican and Italian food bars as well. Apparently there is not enough profit in good salads. Burger King has discontinued them and substituted an almost nutritionless iceberg-lettuce-based one.

Fast food, as a new oligopoly capitalist industrial tradition, is especially devastating to the nutritional status of multicultural/minority groups. Most ethnic traditions contain food combinations that have been the key to their survival. Rice and beans in Mexican food is one example.

Within the African American community, there exist traditions that have used the nutritional (although not preferred by the majority culture) foods in the environment. Aside from heavily fried chicken, African Americans have in their cultural repertoire such healthy foods as collard greens (much more healthy than iceberg lettuce, and containing calcium), chitterlings (which may or may not be "tasty"), and watermelon. All of these foods are nutritionally rich, high in fiber (except chitterlings), and low in cholesterol.

Unfortunately, much of the market for fast food is in low-income and ethnic multicultural/minority communities. And because these communities tend to be heavy consumers of fast food, advertisers target them. As Freedman wrote in the *Wall Street Journal* (1990): "The cycle of poor nutrition is especially intractable in ghetto neighborhoods, in part because of the high-stress environment many must cope with. Junk food can offer an inexpensive means of escape."

Fast-food outlets offer, ironically, some minimal degrees of "status" in ghettos: "Being able to afford a steady diet of fast food shows a customer has some financial muscle." Freedman quotes a Harlem resident who makes $180 per week and spends half of it at fast food restaurants:

> "The atmosphere makes you feel comfortable and relaxed and you don't have to rush," he says, as he admires the hamburger restaurant's shiny floors and picture of George Washington Carver on the wall. Lulled by the soft piped-in music, he nods off for a moment and then adds: "Ain't no hip-hop [music], ain't no profanity. The picture, the plants, the way people keep things neat here, it makes you feel like you're in civilization."
>
> For impoverished consumers, the well-lit restaurants, particularly those representing national chains, are often far more comfortable than home. (Freedman 1990)

But according to Pedro Espada of Soundview Health Center in the South Bronx of New York City, "The industry is public enemy No. 1 to the health of poor, working class people" (Freedman 1990, A1, A6). Freedman notes the specific dangers:

> Sodium intake for an entire day should be no more than 3,000 milligrams for healthy people, says the American Heart Association. Three pieces of Kentucky Fried Chicken's Original Recipe chicken . . . has 1,629 milligrams of sodium . . . and a single biscuit has 655 milligrams. For adults, one Big Mac uses up almost half the maximum daily fat intake of 67 grams: a Burger King Double Whopper with cheese uses up almost all of the fat limit, and has 934 calories.

Because fruits, vegetables, and whole grain breads are more trouble to cook and are more expensive, ghetto nutrition is becoming one of the worst health tragedies in America.

In addition, oligopolized fast food is part of the capitalist industrial process that continues to destroy communal dining traditions. If the family meal is one of the few places where, in our society, children can acquire societal cultural traditions, then fast food contributes to the decline in eating together. Mintz notes that present food marketing and technology aims at "the elimination of the social significance of eating together" (1985, 201). There is some (as yet scanty) research that indicates "that 75% of American families do not take breakfast together. Dinners eaten together are down to three a week . . . and these meals usually last no more than twenty minutes" (Mintz 1985, 205). Among the poor or ethnic minorities, these figures may be worse.

Fast food even destroys restaurant traditions. Often a restaurant meal is a congenial exchange of a group of people who have gone there to enjoy each other's company. This is particularly true in Chinese restaurants where diners share most parts of the meal: "Chipping in for a pizza or splitting a restaurant bill is seen by most people as a more social way of paying for a dinner than dividing the cost according to the number of slices eaten or the price of food and prorated exchange of tip" (Camp 1989, 78). By its practice of having each individual pay for only his or her meal, fast-food dining actively destroys cultural traditions of communal dining.

Fast food offers the illusion, rather than the reality, of individualized choice. The advertising jingle "We do it all for you" in fact refers to a constructed area of the choices offered by a profit-oriented, industrialized food process that relegates the individual's choice to the fast-food company's preplanned selections. Only those that *sell* are considered legitimate "choices."

If one of the elements of high cuisine is surprise (accomplished by the simultaneous lifting of all the covers on diners' platters by the serving staff), fast-food's preplanned, tailorized, preheated choices are the *illusion*, rather than the reality, of "free choice." They actually limit individual choice. What small spirit of adventure and novelty there was in the American cuisine can hardly be found in fast-food eating.[5]

In contrast to the strong cultural tradition of a feast like Thanksgiving, the everyday world of fast food makes a travesty of the traditional strengths of American culture. It destroys adequate nutrition based on a real choice of foods and competition by price and quality. It substitutes preplanned "choice" composed of (usually) high-fat and fatty acid, high-sugar, high-salt meals for the adequate traditions of most ethnic subcultures in American. Through its heavy advertising it convinces (especially the poor) that its offerings are "nutritious." In many ways, fast food is the major force for "nutricide" (Wagoner 1990, 105–108) in America—an ultimately maladaptive "cultural" response to oligopolistic capitalist industrial imperatives.

Summary

In this chapter I have taken a self-conscious and explicit materialist position. As Harris has written: "Cultural materialist strategies are based on the assumption that the biopsychological, environmental, demographic, technological, and political-economic facts exert a powerful influence on the foods that can be produced and consumed by any given human population" (1987, 58).

In this context, food culture is an adaptive response to the material forces in the society. Materialist analysis provides the best explanation of such seeming "irrationalities" as Hindu beef avoidance, Jewish pork proscriptions, and fast-food "cuisine."

But the influence or causation is not all in one direction. Culture allows for the full flowering of human creativity within the limits of successful adaptation to the material world. Harris's position is not "mechanical Marxism" but a statement of the dominant direction of causation, qualified by an appreciation of the creativity and autonomy of cultural response. Harris summarizes his methodology: "While I contend that ecological factors underlie religious definitions . . . I also hold that the effects do not all flow in a single direction. Religiously sanctioned foodways that have become established as the mark of conversion and, as a measure of piety, can also exert a force of their own back upon the ecological and economic conditions which gave rise to them" (Harris 1985, 86). Or one might cite Karl Marx's "The Eighteenth Brumaire of Louis Bonaparte": "Men [and women] make their own history, but they do not make it just as they please; they do not make it under circumstances chosen by themselves, but under circumstances directly encountered, given and transmitted from the past" (Marx and Engels 1962, 247).

There is a place for human will and creativity in food culture, but human beings exercise it in the context of material conditions.

Hindu sacred-cow doctrine, Jewish antipork sentiment, Chinese and Indian cuisine, and American Thanksgiving traditions are all desirable adaptive responses that their respective cultures have made to the material forces in their respective environments. Each now has an autonomous force of its own that we can appreciate as we become involved in the culturally appropriate behavior for each cultural practice.

Fast food, in contrast, is essentially a maladaptive nutricidal response to the imperatives of oligopolistic profit and technological "rationalization in the interests of profit" (Heller 1989). Unless some counterforce in the human or material environment stops the effects of such nutricide, we stand at risk of significantly undermining the effectiveness of contemporary human organisms.

While modern medicine has inordinately expensive operations that can help people with clogged arteries, high blood pressure, and obesity, would it not be more rational to *prevent* such occurrences? If this is not in the interest of oligopolistic kinds of profit, then perhaps it is at least in the interest of the survival of human beings, biotic communities, and the environment.

In chapter 10 we hazard some hunches about the possibilities for change in material food conditions. The impetus comes from the "counterculture," which has evolved an alternative response to the unhealthy one generally provided by the dominant material environment. But first we must examine the international food situation. Included is the pervasive spread of U.S. capital-intensive agriculture—to the detriment of autonomous, indigenous peasant farmers (chapter 8).

Then in Chapter 9 we put world hunger in the perspective of multinational oligopolies that respond to profit rather than need. As multinational fast food circles the globe, so does the destruction of such valuable resources as rain forests now used for increased pasture land.

References

Abrams, Roger. 1984. "Equal Opportunity Eating: A Structural Excursus on Things of the Mouth." In *Ethnic and Regional Foodways in the United States: The Performance of Group Identity*, ed. Linda Keller Brown and Kay Mussell, 19–36. Knoxville, Tennessee: University of Tennessee Press.

Anderson, F.N. 1988. *The Food of China*. New Haven: Yale University Press.

Bryant, Carol A., Anita Courtney, Barbara A. Markesbery, and Kathleen DeWalt. 1985. *The Cultural Feast: An Introduction To Food and Society*. New York: West Publishing Company.

Camp, Charles. 1989. *American Foodways: What, When, and How We Eat in America*. Little Rock, Ark.: August House.

Chu, Grace Zia. 1967. *The Pleasures of Chinese Cooking*. New York: Cornerstone Library.

Counihan, Carole M. 1989. "An Anthropological View of Western Woman's Prodigious Fasting: A Review Essay." *Food and Foodways 3*(4):357–375.

"Countries with the Most McDonalds." 1989. *Lansing State Journal*, November 28, 93.

Cummings, Richard Osborn. 1940. *The American and His Food*. Chicago: University of Chicago Press.

Dean, Karen. 1990. "Reflections on Thanksgiving." Class paper at Aquinas College, Grand Rapids, Mich., November 25.

DeBoer, Diane. 1990. "Reflections on Thanksgiving." Class paper at Aquinas College, Grand Rapids, Mich., November 25.

Deetz, James, and Jay Anderson. 1972. "The Ethnogastronomy of Thanksgiving." *Saturday Review of Science*, November 24, 29–39.

Douglas, Mary. 1966. *Purity and Danger: An Analysis of the Concepts of Pollution and Taboo*. London: Routledge & Kegan Paul.

"Fast Food Could Cost You a Nice Home Life." 1992. *Grand Rapids Press*, March 6, D5.

Freedman, Alix M. 1990a. "Deadly Diet: Amid Ghetto Hunger, Many More Suffer Eating the Wrong Foods." *Wall Street Journal*, December 18, A1, A7.

———. 1990b. "Fast Food Chains Play Central Role in Diet of Inner-City Poor." *Wall Street Journal*, December 19, A1, A6.

———. 1990c. "Poor Selection: An Inner-City Shopper Seeking Healthy Foods Finds Offerings Scant." *Wall Street Journal*, December 20, A1, A5.

Freund, Hugo. 1992. "Celebrating the American Thanksgiving: An Experience Centered Approach to Meaning Formation in a New England Family." Ph.D. dissertation, University of Pennsylvania.

Fu, Charlene L. 1992. "World's McBiggest: Bejing Finds New McDonalds Puzzling but Popular." *Grand Rapids Press*, April 23, B4.

Goodman, Ellen. 1991. "Turkey Is Just Extra: We Feast on Family." *Grand Rapids Press*, November 28, C1.

Goody, Jack. 19984. *Cooking, Cuisine, and Class: A Study in Comparative Sociology*. Cambridge: University Press.

Harris, Marvin. 1985. *Good to Eat: Riddles of Food and Culture*. New York: Simon and Schuster.

Harris, Marvin. 1987. "Foodways: Historical Overview and Theoretical Prolegomenon." In *Food and Evolution: Toward a Theory of Human Food Habits*, ed. Marvin Harris and Eric B. Ross, 57–92. Philadelphia: Temple University Press.

Heller, Peter. 1989a. *Hamburger I, McProfit*. Sydney, Australia: Filmcraft, SPSTV. Videotape.

———. 1989b. *Hamburger II, Jungleburger*. Sydney, Australia: Filmcraft, SPSTV. Videotape.

Hones, Michael, Bruce Giuliano, and Roberta Krel. 1983. *Foodways and Eating Habits: Directions for Research*. Los Angeles: California Folklore Society.

Humphrey, Theodore C., and T. Lin. 1988. *We Gather Together: Food and Festival in American Life*.

Kalcik, Susan. 1984. "Ethnic Foodways in America: Symbol and the Performance of Identity." In *Ethnic and Regional Foodways in the United States: The Performance of Group Identity*, ed. Linda Keller Brown and Kay Mussell, 37–65. Knoxville: University of Tennessee.

Kastin, Aura, and George H. Lewis. 1991. "Celebrating Asparagus: Community and the Rationally Constructed Food Festival." Unpublished paper, University of the Pacific, Stockton, Calif.

Kittler, Pamela Goyan, and Kathryn Sucher. 1989. *Food and Culture in America*. New York: Van Nostrand Rheingold.

Lai, T.C. 1978. *Chinese Food for Thought*. Hong Kong: Don Bosco Printing.

Leidner, Robin. 1991. "What's New About the Service Economy? Lessons From McDonalds." Paper presented at the American Sociological Association, Cincinnati, Ohio, August 23–27.

Lewis, Albert. 1994. Class lecture at Aquinas College, Grand Rapids, Mich., March 16.

Lyman, Stanford. 1989. *The Seven Deadly Sins: Society and Evil*. Dix Hills, N.Y.: General Hall.

Marx, Karl. 1852. "The Eighteenth Brumaire of Louis Bonaparte." In Karl Marx & Frederick Engels, *Selected Works*. Moscow: Foreign Languages Printing House.

McIntosh, Alex. 1992. "Presidential Address." Annual Conference of the Association

for the Study of Food and Society. East Lansing, Michigan, June.

Mennell, Stephen. 1985. *All Manners of Food*. Oxford: Basil Blackwell.

Mintz, Sydney. 1985. *Sweetness and Power: The Place of Sugar in Modern History*. New

York: Viking Penguin.

Murcott, Anne. 1988. "Sociological and Social Anthropological Approaches to Food and

Eating." *World Review of Nutrition and Diet* 55:1–40).

Nair, K.N. 1987. "Animal Protein Consumption and the Sacred Cow Complex in India." In *Food and Evolution: Toward a Theory of Human Food Habits*, ed. Marvin Harris and Eric B. Ross, 445–454. Philadelphia: Temple University Press.

Pelto, Pertti J., and Gretal H. Pelto. 1987. "Culture, Nutrition and Health." In *The Anthropology of Medicine*, ed. Ross Romanacci, D.F. Moerman, and L.R. Tannered, 173–200. New York: Praeger.

Prakash, Om. 1961. *Food and Drinks in Ancient India*. New Delhi. n.p.

Sahni, Julie. 1980. *Classic Indian Cooking*. New York: Morrow.

Sanjur, Diva. 1982. *Social and Cultural Perspectives in Nutrition*. Englewood Cliffs, N.J.: Prentice Hall.

Sapp, Robin. 1990. "Reflections on Thanksgiving." Class paper at Aquinas College, Grand Rapids, Mich.

Simons, Frederick J. 1980. "The Sacred Cow and the Constitution of India." In *Food, Ecology and Culture: Readings in the Anthropology of Dieting Practices*, ed. J.R.K. Robson, 119–124. New York: Gordon and Breach Science Publishers.

Tannahill, Reay. 1984. *Food in History*. New York: Stein and Day.

Vaz, Rozanne. 1990. "My Views on Thanksgiving." Class paper at Aquinas College, Grand Rapids, Mich., November.

Wagoner, Michael. 1990. "Nutricide." *Z Magazine*, July/August, 105–108.

Wieroky, Jessica. 1990. "Thanksgiving." Class paper at Aquinas College. Grand Rapids, Mich., November.

Wootan, Marge, and Bonnie Liebman. 1993. "The Great Train Wreck." *Nutrition Action Newsletter*, November, 10–12.

Yoder, Don. 1972. "Folk Cookery." In *Folklore and Folk Life*, ed. Richard M. Dorson. Chicago: University of Chicago Press.

Notes

[1] 1.One fantasy way was provided with pigs in the popular film *Mad Max, Beyond Thunderdome*. All the pigs were kept underground to collect the excreted methane that powered the society above ground.

[2] 2.In fact, the Phoenicians, Egyptians, and Babylonians were also avoiders of pigs. It is only later, with the Diaspora, that pork avoidance became a specifically Jewish food marker.

[3] 3.There are, of course, other aspects of the Jewish food proscriptions that may not be explained with such facility. For instance, from where comes the proscription against mixing meat and milk? And what is the origin of the need to drain all the blood from the slaughtered animal? These are interesting research questions for social scientists studying food.

[4] 4.Freund (1992, 37) quotes McMahon concerning a latent function of preparing much too much food: "Much of the preparation for the Thanksgiving table actually served the [farm] through winter."

[5] 5.By building their offerings around beef, fast-food oligopolies are also major contributors of the despoliation and poverty of much of the developing world. Because beef brings higher profits, multinational corporations replace the cultivation of traditional nutritious crops with grazing land. They even take rain forests. And, by their capital-intensive methods of industrialized beef farming, they throw the previously self-sufficient peasants into the ranks of the poor and hungry in the cities of the developing world. See chapters 8 and 9.

Chapter 7

Food and the Social Order

INSTRUCTIONS: Match the following food events with their appropriate meaning.

Food Event

A home cooked meal by candlelight
Sunday Brunch
Saturday Lunch
A cup of coffee
Dinner at an elite restaurant
Lunch at McDonald's
Co-cooking at his or her house
Having a drink at a "nice" bar

Meaning

A first meeting
A casual get together
Celebration of an ongoing, long term relationship
A night of romance
A business meeting
Getting to know you better
A first date
A routine meal in an extended relationship

How did you come out in this exercise? Does everyone in the class agree? What is the point of the exercise?

Most of us would agree that the degree of luxury and care put into the meal mirrors or reflects the quality of the relationship between two people in a relationship: "Small wonder that the word 'companion,' which connotes friendship, warmth, and security, stems from French and Latin words meaning 'one who eats bread with another'" ("Breakdown of the Family Meal," 1991, 3). Or, as Kahn observes, "the act of eating 'together' . . . is highly symbolic. . . . Behaviorally, 'to eat together' indicates openness and trust of one's fellow human beings" (1989, 134).

One point made by the opening exercise is the parallel relationship between the quality of the meal and the degree of intimacy of the friendship/romance. Implicit in the exercise is an ideal-typical progression that begins with a cup of coffee (or tea, or a Coke) and progresses through a romantic, candle-light dinner. A further stage might be the routinization that comes from a co-cooked meal. Again, the quality of the relationship would be symbolized by the amount of time, care, and money invested in the meal.

This chapter focuses on food as it transforms biological hunger into social indicators of status. In the first instance, food is an indicator of the closeness of interpersonal and group relationships. In the second, food indicates a level of social stratification or social class.

Both perspectives take place within the idealist/materialist dialectic. Food has a symbolic meaning that often evidences the material relationship between persons. But one can consciously manipulate the meanings of food to communicate desired meanings and desired degrees of closeness. Nothing indicates this better than the use of elaborate food prepared with great care in a seduction experience.

From a materialist perspective, food is often an indicator of status/class differences. People at different levels in the society eat different kinds of food. Yet, within the limits of material supports, various groups exercise creativity and art in the preparation and presentation of food. The cuisines of a variety of ethnic and racial groups are the result.

Food in Social Interaction

Generally the sharing of food has to do with social boundaries. When we examine class, gender, and race (in sociological jargon, stratification), we see food used as a mechanism for delineating between social groups. In social interaction, it functions to unite the "in" group that is sharing it.

The sharing of food is about the reinforcing of social intimacy. As Moore argues, "Generally we humans accept food most readily from our friends and fear the food of strangers" (1957, 78). Food, like the Christian Last Supper, bonds its participants together. The Eucharist has celebrated unity within the Christian community in the present and through the ages. By eating the communion bread and drinking the wine, Christians affirm their unity with the church and with each other.

In contrast, we generally fear the food of enemies and strangers. One recalls the ethnocentric joke about taking one's dog to a Chinese restaurant, leaving it in the entry, not finding it on the end of the meal, and wondering whether, in fact, one had some of it for dinner. By telling this joke one "puts down" Chinese food practices (which include eating dogs and cats). And one affirms one's (ethnocentric) rightness, in decrying the "abuse" of what Westerners consider to be household pets.

A more extreme way one degrades strangers is the attribution of cannibalism. Tribal societies often bred fear in their members by threatening that their enemy would "eat you."

More famous is the manner in which Romans of the first four centuries A.D. regarded Christians. At that time, Christianity was a small minority sect in the midst of the dominance of the Roman pantheon. In classic food terms, the Roman establishment fixed on the Christian Eucharist as inhuman. Did not the Christians partake of the *body and blood* of Jesus Christ? Did they not *eat* their God? Was this not cannibalism? Who could deny it? By decrying this Christian cannibalism, the Roman establishment was affirming the essential rationality of the Roman pantheon and social system (including its eating habits) and relegating this small, strange sect to an outcast position.

The meal is the food event that focuses the analysis of the functions of eating together. As Seymour writes, "in our culture, people eat in family groups comprising parents and children" (1983, 3). And "the serving of a hot meal indicates intimacy" (Seymour 1983, 4).

If eating together is a mark of unity in the group, one must also examine the manner in which that unity is actively reproduced. In sociology, this process is deemed *socialization.* Unlike the everyday usage of the same term (circulating at a cocktail party), sociologists use socialization to denote the process of becoming "a functioning member of society" (Seymour 1983, 4). Children learn the norms, values, ideas, beliefs, and ways of acting appropriate to their society, subsociety or subculture.

The family meal has often been the centerpiece of this learning experience. In "The Social Functions of the Meal," Seymour writes:

> During mealtimes the child learns how to eat, what to eat, what kind of food is available to whom, how to evaluate food—he hears adults comment about the food, its plenty or scarcity, tastiness, monotony, its effects on health and so on—in other words he is introduced to the society's stock of knowledge about food. The child also learns about appropriate behavior around eating and through this he learns about appropriate behavior in other spheres. He...learns about manners and etiquette, and it is one of the first social occasions at which he is required to learn self control, and exhibit proper social behavior. (1983, 6).

Seymour describes mealtimes as an opportunity for the child to learn about and enact new social roles: "Mealtimes provide an opportunity for the child to learn more about the roles available in the social group and how to play them; later children play at these roles by, for example, giving their dolls tea parties" (1983, 6).

Mealtimes provide an opportunity to act out the roles with which society presents us: "A meal provides the opportunity for everyone to display their control of their role performance" (Seymour 1983, 5). One can be "host," "guest," "server," "rebellious adolescent," "family member," or a host of other societally sanctioned roles.

Even choice of food may be an opportunity for role display. Health-conscious eaters may take the skin off a served chicken. This indicates their knowledge of the high fat in chicken skin and their resolve, based on contemporary medical knowledge, to remain free of arterial diseases. There is no more cholesterol in the skin of a chicken than in the lean meat—more fat, but no more cholesterol. Cholesterol distributes itself evenly in all the parts of the same animal!

In the same manner, refusing food partakes of certain roles. Vegetarians must, by definition, refuse meat dishes. In the context of food's conveyance of intimacy and love, this refusal is usually socially problematic. How does one accept the love but reject meat (perhaps the strongest status symbol)? Personally, I have attempted to alert people in advance. When that fails, some meals have been rather sparse by guest dinner standards.

Childhood and adolescent rebellion may be enacted most poignantly at the dinner table. It is usual for parents to encounter childhood rebellion focused about a child who will not eat the food presented. Books on childrearing abound with suggestions for feeding the child or, at least, making sure the child does not starve.

Anorectics take this refusal further. Unlike vegetarians (who will eat their desired food), anorectics often refuse all (or most) food. While this may be treated as a medical problem, one can also argue that it is adolescent rebellion taking a modern form. A teenager refusing to eat a mother's (or father's) meal can manipulate the affections and anger of any seemingly normal family.

Additional role performance can be found in the enactment of food sex roles. In contemporary Western society, it is traditional for the woman to serve and the man to carve. And the status of the various participants can often be judged by the size of their portions. Throughout the world, women and female children usually get lesser quality and quantity.

For the most part, it is mealtime when the family performs its unity. For quality growth, "children need the structure and discipline involved in getting the family together at mealtimes in order to feel secure. . . . Indeed, psychologists have found that high-quality feeding interactions between parents and young children heighten the children's cognitive ... abilities along with their linguistic competence in later years" ("Breakdown of the Family Meal" 1991, 4).

Unfortunately, the sharing of common meals has declined significantly in the United States. Beginning before the 1929 Depression, Robert and Helen Lynd wrote: "Mealtime as family reunion time was taken for granted a generation ago; under the decentralizing pull of a more highly diversified and organized leisure . . . there is arising a conscious effort to 'save mealtimes, at last, for the family'" (1929, 153).

"Morning, noon and evening meals eaten within the home are still the rule" (Lynd 1929, 154). However, as McIntosh (1992) writes, "contemporary National Food Consumption Survey data indicate that dual working parents and their families eat out more frequently." Industrial work patterns lead to convenience foods, microwave ovens, and fast-food restaurants. All contribute to the demise of the family meal.

> The reality is that many American families do not eat together today. Consider that in a recent national poll of 555 parents with children under 18, it was discovered that one in five families had not had dinner together the previous evening. Another poll, surveying 1,000 people in the Los Angeles area, revealed that one in three households did not eat regularly as a family. ("Breakdown of the Family Meal" 1991, 3).

This situation may correlate with the decline of the African American family, especially at the lowest levels of income.

Perhaps the time when the unifying function of the family meal is performed best is a holiday season. Meals like Thanksgiving (chapter 6), Christmas, and Easter serve to unite existing families and solidify ties with past families. Recipes for turkey, stuffing, cranberry, and Christmas cookies fill the recipe boxes of most (female) family cooks. One may often send food to distant family members as a reminder of family solidarity. As Ellen Goodman (1991) writes: "Turkey is just an extra; we feast on family."

This chapter began with a brief matching exercise with romantic intimacy and meal type. Some thinkers consider this connection even more direct. Goody quotes Lévi Strauss "who has insisted upon the identification of copulation and eating [where] two separate but complementary units unite. In India, the word for eating, '*di*,' is frequently used for sex, and covers much of the semantic field of the word 'enjoy' in English (1984, 114). Similarly, in China of the Ch'ing period, as Spence points out, "it has often been said that the vocabularies of food and lust overlapped and blended into a language of sensuality" (Goody 1984, 114).

It is an axiom of Western culture that "the way to a man's heart is through his stomach." Women garner the attention of men through preparing meals for them (and sometimes vice versa). In the historical novel *Roots*, the chief cook, Bell, got the attention of hero Kunte Kinte with tidbits from the kitchen of the slave master. As the relationship progressed, she prepared more and more elaborate meals for him. Finally, they made love for the first time at her candlelight dinner. This is certainly not a romance that fits the white stereotypes. Nevertheless, Alex Haley's historical novel stays as true to historical accuracy as possible. Yuppies are not the only group that knows the value of the candlelight dinner.

In polygamous societies it is the woman who prepares the food who has the sexual relationship that night with the husband. Women who want to keep the affections of the husband must seek to prepare a good meal for him.

If the use of food contributes to solidarity within the family and between romantic/sexual partners, it may also be used as a criterion by which to judge others. Just as "out-groups" are often stereotyped as cannibals, those people(s) who share food are often held in high esteem. In Christian literature, Jesus is portrayed in the parable of the loaves as sharing bread with a large gathering of people. Similarly, in her study of Melanesian society, Kahn found that "generosity is accorded the highest social value. Any food that appears in public must be shared with as many people as see it" (1989, 41). She quotes the Wamirians: "We are not like white people, we share our food" (Kahn 1989, 41). Could it be that the sharing of food as a social norm is, perhaps, what less "developed" cultures may have to teach the affluent West? Some contend that the practice of sharing food was the "insurance policy" that explained the cooperative success of the original human societies—hunters and gatherers.

Contemporary public eating practice continues to demonstrate the boundaries of groups. Eating clubs, fraternities, and sororities limit dining together to their members— thereby reinforcing group unity. In mass dining halls, cliques will eat together. In many college cafeterias, the athletic teams eat together (there may even be a special table). And sometimes racial, ethnic, or religious minorities bolster their unity (and generate resentment) by limiting partners with whom they dine in public.

Food has the potential for establishing unity within the group that shares and eats it. The issues of with whom one does *not* eat form the material for an analysis of stratification and food.

Food and Social Stratification

> Tastes in food . . . are socially shaped, and the major forces which have shaped them are reli-

> gious, classes and nations. In European history, religion has been a relatively weak influence on food, class overwhelmingly the strongest. People have always used food in their attempts to climb the social ladder themselves, and to push other people down the ladder. Today, it is possible to speak of elite or highbrow food, popular or mass cooking, folk cookery, even junk food. Inevitably these notions provoke controversies parallel to those between protagonists of high culture, popular culture and folk tradition (Mennel 1985, 17).

Just as eating common food together serves to unite groups and reproduce their common culture, so the eating of different foods serves to exclude groups. In the early stages of the American civil rights movement, the targets for integration were the restaurants from which whites wanted to exclude African Americans (Lowenberg 1974, 150). Eating together expresses equality.

Historically, the type of food eaten has been one primary manner in which the upper class has distinguished itself from the masses (peasants, serfs, beggars, and later the working and middle class). As Gofton writes: "Historical studies reveal that differences in food consumption practices are, of course, largely founded on inequalities of economic and political power" (1986, 129). According to Mennel: "Higher social circles have repeatedly used food as one of the many means of distinguishing themselves from lower rising classes" (1985, 332). The use of food reflects the class-status system.

Generally the food features that distinguish the upper class relate to type, amount, and manner in which food is eaten. The mark of the upper class throughout history has been its conspicuous consumption of massive quantities of meat. The *potlatch* was one form of this (see chapter 6).

Many know Roman society by its food (as well as sexual) orgies among the upper classes. The social pressure to gorge was so great in Roman society that many buildings contained an adjoining "vomitorium" for the specific function of regurgitating what one has just eaten so that one could then eat more. Such practices "led to a series of sumptuary laws through which an attempt was made to control the expenditure on food and to limit the extent of conspicuous consumption" (Goody 1984, 103).

Goody quotes Elias's description of the food practices of the upper class of medieval society where

> . . . the dead animal or large parts of it were often brought whole to the table. "Not only the whole fish and whole birds (sometimes with their feathers) but also whole rabbits, lambs, and quarters of veal appear at the table, not to mention the larger venison or the pigs and oxen on the spit" (Elias 1978, 11) . . . . The diet of the Northern German court showed its resident members to consume two pounds of meat a day in addition to large quantities of venison, birds and fish. (Goody 1984, 139).

In this context, the religious community set itself off from the secular elite by denying this affluence in food. Monasteries practiced various degrees of food regulation. These included restricted consumption and various types of fasting. For the Christian church, "a constant theme of its ideology, especially as expressed by St. Augustine, was an aversion to 'gluttony' as one of the seven deadly sins" (Goody 1984, 139). "Christian moralists saw in elaborate foods and eating ceremonials a way the devil acquired disciples" (Goody 1984, 139).

For a limited time, while sugar was both scarce and expensive, it functioned like many spices available only to the elite. Sugar came to Europe about A.D. 1100 as the result of cultural diffusion associated with the Crusades. For two or three centuries, it was the almost exclusive possession of the rich. It appears in elite recipes as well as medical tracts (it was used both as a medicine and as a dentifrice!). As Mintz writes: "Differences in quantity and form of consumption [of sugar] expressed social and economic differences within the national population" (1985, 79).

Spices were the method by which the rich created variety in taste. It was their prerogative because they were able to afford expensive imported spices, to have a much greater range of flavor (and preservation) than the poorer classes: "The somewhat indiscriminate use of sugar in flesh, fish, vegetable, and other dishes is evidence that sugar was regarded as a spice" (Mintz 1985, 84).[1]

In contrast, the food most characteristic of the common or poor people's diet in Western society has been bread. The price of grain often dictated the health of the working people. Of course, the bread was whole grained, not the white, nutritionless wonder of much of contemporary industrial society's working class.

The rich also distinguished themselves from the mass of society by their style of eating. This embraced both manners and location. Gofton notes that it is common for archaeologists to correlate physical form with social relationships (1986, 137). And Mennel documents the withdrawal of the lord from the common hall to a separate dining chamber as evidence of the increased power of elite/mass class divisions (1985, 57).

Similarly, the elite often portrayed itself with gentler sensitivities. The civilizational process traced a distancing from the baser aspects of society. Thus the original center table carving of the Middle Ages gave way to carving at the side of the table (and, later, precarving). The fork and spoon were added to supplement the knife as an eating device (Braudel 1979, 203, 205–206).

Industrial (capitalist) society brought about the second major change in class structure in human civilization (the first was the separation of elite and mass in food-growing societies from the approximate egalitarianism of hunting and gathering or foraging societies). The new elements in the class structure were the rise of the bourgeois (the middle class of urban burgehrs) to industrial capitalists and the creation of an industrial working class in the cities parallel to the peasantry of the countryside. Changes in food consumption content, style, and place parallel these refigurations of the class structure. Because newer and lower classes usually imitate older and higher classes, one can trace the changes in class eating content through bread and sugar.

In medieval society, the bread the upper class ate was significantly whiter. The working class ate darker bread. With the invention of steel and later porcelain rollers, industrial milling perfected the process of creating completely white bread. From our current perspective, the rich were eating an almost nutritionless product, while the dark bread of the poor provided many of their basic nutritional needs (especially B vitamins and bran). The rich, of course, had many other sources of nutrition.

As the Industrial Revolution continued, mass production techniques replaced home baking. White bread became the status standard of the working class. It is only with the recent popularity of holistic eating that darker bread reappears at the upper levels of the social scale (Mennel 1985, 303).

Such a situation causes Mennel to project an Adorno critique of food in society whereby he supposes that the working class "had been 'forcibly retarded' by the food industry" (1985, 320) and, as in George Orwell's 1937 contention, "now rejects good food almost automatically" (Orwell 1937, 89).

In a similar manner, Sydney Mintz has traced the coincidence of profit motive and working-class imitation of upper-class eating habits about sugar. As sugar (and tea) became part of the rationale for the expansion of the British Empire in the eighteenth century, they came to fill the pockets of entrepreneurs who invested in them. At the same time, because consumption of these items had previously been limited to the upper and bourgeois classes, they became desirable tastes to emulate.

Unfortunately, from a nutritional perspective, the substitution of tea and sugar (which provided a "lift") for beer (which was highly nutritious then) and dark bread brought a precipitous decline in the quality of working-class nutrition. Because tea has no nutritional value (the first completely "junk food") and sugar only empty calories, working-class diets lacked protein, complex carbohydrates, vitamins, minerals, and fiber. Nutritional limitation (especially for women and children) became "a kind of socially legitimized population control" (Mintz 1985, 149).

Was this depletion of the working-class diet the result of a conscious conspiracy of the industries that profited from it? In their quest for profit, capitalist industrialists laid the foundation for the decline of the nutritional value of the Western diet. It is "too high in sugar, salt and fat, deficient in fibre [and] is a major cause of degenerative diseases in western societies" (Mennel 1986, 321).

While it is true that supply and demand had a major hand in shaping this process, it is also true that many capitalist interests came to have a major stake in poor nutrition. As Mennel writes: "These things set in motion a compelling, self-reinforcing process which on the one hand shaped popular taste, and on the other hand, established vested interests which the industry has been known to defend ruthlessly even in the light of more recent nutritional knowledge" (1986, 321).

Ironically, the rural poor had more resources to feed themselves adequately than those working classes who labored in urban factories. Because food was produced in the countryside, people often grew their own vegetables and had some access to the less desirable (but highly nutritious) parts of animals.

Often cited as one of the most resourceful and healthy eating traditions is that of African Americans. Genovese documents the manner in which American slaves developed the tradition of drinking the "pot likker" (the water in which vegetables were cooked) (1976, 548). Only later nutritional discoveries documented the value of this vitamin-laden liquid. Whites of that era disdained it. But most of them had a wider range of nutritional options available to them.

In addition, mixing West African tradition with available foodstuffs, the African American slaves combined beans, cabbage, and hog jowl with squash and okra to make a most nutritional mix (Genovese 1976, 548). They mixed poke salad with ashcake or hoecake and clabber, buttermilk and sweet milk. This mix contained not only superb nutrition but a degree of tastiness often missing from the slavemaster's bland diet. Slaves faced the monotony of meager grains with a creativity and care that created a healthy and tasty cuisine.

Both in slavery and immediately afterward, poor southern African Americans utilized the animals that whites did not eat (raccoon and possum) as well as the "undesirable" parts of animals whites did eat. In the latter case, chitterlings (pig intestines), pigs' feet, and ribs were the results that have come down to the present era as African American cuisine. African Americans provide one example of a poor people's rural culture that sometimes ate reasonably well in the face of oppression. This was partly through being near to the sources of food and partly to West African traditions combined with culinary creativity. "Soul food" is the result.

Nevertheless, it is most difficult to maintain healthy eating traditions in the face of the combined power of contemporary fast-food marketing and the kind of advertising that preys on lower-income desires to emulate statuses above them. If the upper class uses food to set itself apart, food advertising can utilize emulative desires in selling nutritional garbage to the poor.

Because of the cleanliness, attractiveness, speed, relatively low cost, and willingness to locate in a variety of ghetto and low-income neighborhoods (see chapter 6), the fast-food establishment is at present involved in a process that parallels the introduction of tea and sugar as the replacement for beer, bread, and butter.

Just as the African American community owes a lot of its successful survival to the nutritional combinations it developed in slavery and from Africa, so the Hispanic community has a nutritionally sound cuisine. The traditional rice, beans and corn combine vegetable protein in such a way as to make it complete (Sanjur 1982; Kittler and Sucher 1989).

But these cultural traditions (originally adaptations to the existing food situations) do not fully succeed in resisting the onslaught of modern capitalist industrial advertising. Preying on the traditional desire of low-income people for meat (perhaps an imitation of the 1950s upper- and middle-class barbecue), the fast-food industry has moved in. In a *Wall Street Journal* series on ghetto nutrition, Alix Freedman quotes a Hispanic youth as saying, "'Everyone is tired of their mother's food—rice and beans over and over. I wanted to live the life of a man. Fast food gets you status and respect'" (Freedman 1991, 1).

Between the necessity for ghetto moms and dads to work, the bias of many state welfare laws toward single-parent families, and the effectiveness of highly paid social scientists who work for the advertising industry, the low-income ghetto becomes just one more "market" for the fast food oligopolies. "Style" overpowers nutrition, and many ghetto people become malnourished and sick. In Spike Lee's 1991 *Jungle Fever* the crack addict brother seems to live on candy bars.

Hunger and Homelessness

At the bottom of the present class hierarchy are America's increasingly populous homeless. As the Harvard Physician Task Force on Hunger in America reported in 1987:

> After more than two years of field investigations in the diverse regions of the nation, we have reached a shocking conclusion. At least twenty million Americans were going hungry at some time each month . . . , a level of human misery reminiscent of the Great Depression. Equally distressing, we had uncovered evidence that this hunger was a man-made epidemic created and spread by government policies. (Wilber 1988, 51)

Ultimately, the problem of hunger amid plenty is one of world proportions. In the midst of massive consumption of luxury items (meat, elite fruits), there is also massive poverty throughout the world. This subject occupies chapters 8 and 9. But such widespread hunger in the richest country on earth is sufficiently shocking to merit some discussion here.

Neither hunger nor poverty is new to the United States. What is disconcerting are our attempts to camouflage it and stonewall inquiries into its extent. Macroscopically, both at a world and a country level, the problem is one of the provision of aid (in the form of surpluses) within the context of the capitalist oligopolistic market system. Whether in a developing world famine or in the U.S. Great Depression, pushing free food into a market economy undercuts what are considered the legitimate profits of people who sell food. If the government provides free bread, what happens to the price of bread in the general market? The price goes down precipitously.

This problem, in the U.S. context, was nowhere better illustrated than by the activities of government during the Great Depression. In *Breadlines Knee-deep in Wheat*, sociologist/historian Janet Poppendieck documented the existence of large surpluses of wheat (as well as milk and pigs) in the midst of 25 percent unemployment and widespread hunger. As she writes:

> While oranges were being soaked with kerosene to prevent their consumption in California, whole communities in Appalachia were living on dandelions and wild greens. Corn was so cheap that it was being burned for fuel in country courthouses in Iowa, but large number of cows, sheep, and horses were starving to death in the drought-stricken Northwest. Dairies were pouring unsalable milk down the sewers, while unemployed parents longed to provide even a pint a week for their growing children (1986, xi–xii).

The response of the Hoover administration paralleled that of Reagan and Bush. As the Lynds detailed in *Middletown in Transition* (1937), the government responded with a call for voluntary programs similar to "a thousand points of light," Churches and health organizations (the Red Cross) became distributors of contributed food. Breadlines and soup kitchens became familiar sites in America's cities. Yet the Depression's poverty and hunger continued unabated.

In this context, Franklin Roosevelt assumed the presidency in 1932 with a commitment to involve the government in food and poverty relief programs. Much of the United States' basic social welfare system evolved from these programs.

> In food, the mechanism of aid was the Federal Surplus Relief Corporation. It was seen by its creators as a temporary emergency measure to transfer agricultural surpluses to the unemployed until such time as the New Deal could bring about recovery. . . . In fact, the policies . . . established by FSRC endured for decades and continue even now to influence domestic food assistance policy and politics. (Poppendieck 1986:129).

Section 32 of Public Law 320 was the vehicle of authorization: "By mid 1936 . . . more than two thousand local, district and state work projects for the distribution of surplus commodities were in preparation or had been approved" (Poppendieck 1986, 198).

With the transfer of food assistance programs to the Department of Agriculture, the United States created the beginning of its food stamp policy between 1939 and 1941. It initially served 4 million people (Poppendieck 1986, 241). And 1949 marked the passage of PL 480 to distribute additional surplus food to developing nations.

The idealism of the 1960s was reflected in the passage of childhood nutrition programs and extensions of the food stamp program. With America's rediscovery of poverty amid affluence, the War on Poverty produced the most socially useful legislation in the history of the United States. In that decade the food programs originated that, until recent administrations, formed the food safety net for the hungry. One was the WIC (Women, Infants and Children) program to provide pregnant women and mothers of small children with adequate health care and food. School lunch and senior nutrition programs were others.

Since the late 1980s, America has again been experiencing a crisis of domestic hunger in the midst of plenty. A large number of scholars have documented the increases in income of the upper strata of American Society: "In 1986, the richest fifth of American families received 43.7% of national family income—the highest percentage ever recorded" ("The Numbers: Hunger and Poverty" 1988, 31). Skylar writes that "census data indicate that the gaps between both the rich and the poor and the rich and the middle class are wider now than at any other time since the end of World War II" (1991, 10).

At the same time, America is experiencing a wave of homelessness not seen since the Great Depression. Twenty percent of Americans experience hunger once a month or more. Forty-three percent of African American children and

37 percent of Hispanic children live in poverty and experience hunger ("The Numbers: Hunger and Poverty" 1988, 31).

Malutrition affects almost 500,000 American children each month according to the Physician Task Force On Hunger in America ("Hunger Numbers" 1989, 16). This same group argues that "from 1982 to 1985, $5 billion was cut from the four basic child nutrition programs: school lunches, school breakfasts, child care food and summer food programs" ("Hunger Numbers" 1989, 14).

According to the *Grand Rapids Press*, in 1991, 89,055 schools took part in the National School Lunch programs but fewer than half offered breakfast; 24.9 million children ate school lunches in November 1991, but only 4.9 million got breakfast.

Standing behind hunger is the American institutional structure that has led to an increase in poverty and homelessness. Another major factor has been the "deinstitutionalization" of a large number of American mental health patients. In the context of "therapeutic innovation" (a variety of outpatient treatment plans and group living), severely disturbed former patients are now saving money for the government by living on the streets and eating, when they can, at soup kitchens and garbage dumps. Also contributing was the virtual elimination by the Reagan/Bush administrations of the public and subsidized housing programs. As housing became less affordable by low-income people, more and more were forced to move to the street as the only economically viable alternative. With unemployment very high (especially for minority youth), and with an inadequate minimum wage, even people with jobs cannot generate the kind of income thatt would keep them in adequate shelter. Entire families are showing up in soup kitchens and in shelters. In affluent California, "2.3 million people rely on emergency food charity every month" (Brown 1988, 12).

Once again, as in the Great Depression, the response of recent administrations has been one of patchwork programs and platitudes about voluntarism. In spite of the precedents set in the Depression and in the 1960s, Republican administrations systematically avoided serious governmental action to alleviate the plight of America's increasingly hungry and homeless. The 1987 Stewart B. McKinney Homelessness Assistance Act provided a shotgun of inadequate and piecemeal programs. Of the $6 million authorized, half was not used. The Clinton administration has yet to improve greatly on the "efforts" of its immediate predecessors.

In spite of this lack of support, many volunteers struggle in the finest traditions of American private efforts. Their helping efforts in the area of food have taken the form of food contributions, gleaning, and direct feeding programs. One source of aid is the restaurant business. The year 1984 marked the creation of Share Our Strength, an organization of 2,000 chefs and 10,000 food service professionals (Barker 1990, 8). This organization "fights hunger by raising money for food banks, perishable food programs, and homeless shelter programs" (Barker 1990, 8). In addition, such organizations as the Pikes Peak Chapter of the Colorado Restaurant Association helps the Salvation Army with the collection and distribution of prepared food from "restaurants, caterers, bakeries and grocery stores in Colorado Springs" (Barker 1990, 9). In Phoenix, an organization called Restaurants In Partnership, Inc., was launched in 1989 for the purpose of linking restaurants across the country with social agencies to provide a voucher system (Barker 1990, 9).

A second helpful effort comes in the form of gleaning operations in a large number of organizations. Many foods never reach the grocery shelves because of dented boxes or cans, overripe fruit, asymmetric shapes, or odd sizes. Newspaper columnist Colman McCarthy cites the Maine Potato Project which links potato producers in twenty-four states (1990, A12). As one of the most nutritious vegetables, containing vitamins, protein, and fiber but no cholesterol or fat, this potato project has been helping supply the soup kitchens of America.

In western Michigan, the Second Harvest Gleaners Food Bank, begun by the Reverend Don Eddy and now led by John Arnold, distributed 7 million pounds of food to charity in 1989. With a staff of volunteers and a computer, it serves a forty-county area in western Michigan (Harger 1991, C3).

Soup kitchens again abound in America. In Grand Rapids, Michigan, the two most prominent are Capital Lunch ("God's Kitchen") and Degage Coffee House. The former provides free lunches every day from donations; unlike state-sponsored welfare organizations, it has eliminated the degrading bureaucracy. One need merely show up between noon and 2:00 p.m. at their offices on South Division Street downtown and eat. Director Barbara Raaymakers provides a material and spiritual presence that organizes a massive volunteer support system, raises funds, and settles disagreements among people eating.

For evening meals at the Degage Coffee House, a staff of volunteers (many of them college students and church members) makes possible a very-low-cost meal (if necessary it can be free to those who help clean up) to the homeless and people in poverty. Such organizations have become common on the American urbanscape.

More politically active and national in scope is Food Not Bombs. It has organizations in Boston, Washington D.C., Houston, Atlanta, Minneapolis, New York City, San Francisco, Sacramento, and Long Beach, California (Vitale 1990, 21). Unfortunately, San Francisco has sought to curtail its free vegetarian food distribution project by arresting its personnel—they compete with the market. As Alex Vitale writes: "Food Not Bombs is working to create and encourage alternative institutions of food distribution based on human production and distribution methods and not corporate profits" (1990, 22).[2]

As in the early stages of the Great Depression, those who profited from the corporate system sought to stop any questioning or undercutting of the profit structure. If free food is distributed, who will buy the junk food huckstered by America's multinational corporation food giants?

This problem is fundamental to the world corporation system based on profit. It explains a great deal of the ineffectiveness of hunger relief efforts throughout much of the developing world. But it has been effectively overcome in such socialist states as China, Cuba, Sandinistan Nicaragua, North Korea, and the Indian state of Karala (see chapter 10).

Summary

One can use food to indicate either unity or division. Within groups common food consumption creates and reinforces unity. Most important, the family meal has served socialization functions as well as providing an emotional haven from the public world. Food marks the boundary between in groups and out groups.

Food is greatly bound up with romance. People use a variety of formats and foods to signify their estimation of the importance and depth of a relationship. And there is some psychological support of the equation of food and sexuality.

But people who eat different foods often fall at different levels of the social stratification hierarchy. The rich often use their influence for more variety and more, in whatever terms the era conceives it, "healthy" food.

Generally, the middle of the social spectrum emulates the rich of previous generations. At present, beef consumption seems to fulfill their cultural need for status imitation. Yet with the rich now beginning to eat like the world's poor, one wonders what to expect next from America's middle classes.

The poor (who can afford to) emulate the middle classes. Aside from economies of scale, one might well attribute the salability of McDonald's and the other oligopolies of hamburger to status emulation. Not only the food but also the atmosphere stands in contrast to many of the discomforts of low-income life.

At the bottom of the stratification ladder are the homeless hungry. Increasingly a casualty of an economic system that refuses to compromise its commitment to corporate profit, the homeless hungry eat the leftovers of the consumption patterns of twentieth-century America.

Thousands of volunteers staff the private feeding programs that seek to keep America's homeless and hungry from an even less desirable existence. In contrast, a publicly "compassionate" but institutionally uncaring government refuses to support the necessary governmental programs that could begin to alter this situation in a serious manner. Surplus food distribution has been curtailed. Free seeds are no longer dispersed. New low-wage service jobs (often in fast-food restaurants) do not pay enough on which to feed a family. And there are no programs to make the *institutional* changes necessary in order to provide everyone in the richest nation on earth with adequate food and nutrition.

This is not a problem peculiar to the United States. In the context of a world capitalist system, all but the socialist world permit beggary, destitution, and famine to coexist with private affluence and oligopolistic profit. Chapter 9 examine this situation on a world scale. But first we need to explore food production—the oligopolized technology of world agriculture.

References

Barker, Lori. 1990. "Community Spirit: Restaurants Reach Out to Neighbors in Need." *Restaurant Business*, September 20, 8–12.

Braudel, Fernand. 1979. *The Structures of Everyday Life: Civilization and Capitalism, 15th–18th Century*, vol. 1. New York: Harper & Row.

"The Breakdown of the Family Meal." 1991. *Tufts University Diet and Nutrition Letter* 9(5):3–5.

Brown, J. Larry. 1988. "The Paradox of Hunger in a Rich America." *Food Monitor 43*:12–14.

Cosman, M.P. 1976. *Fabulous Feasts: Medieval Cookery and Ceremony.* New York: Braziller.

Elias, N. 1978. *The Civilizing Process*. Oxford: Basil Blackwell.

Fersh, Robert J. 1988. "Eliminating Hunger: The Agenda for Food Assistance." *Food Monitor 43*:22–24.

Fieldhouse, Paul. 1986. *Food and Nutrition: Customs and Culture*. New York: Croom Helm

Freedman, Alix M. 1990a. "Fast Food Chains Play Central Part in Diet of Inner-City Poor." *Wall Street Journal*, December 19, A1.

———. 1990b. "An Inner-City Shopper Seeking Health Food Finds Offerings Scant." *Wall Street Journal*, A1, A8.

Genovese, Eugene D. 1976. *Roll Jordan Roll*. New York: Vintage Books.

"God's Kitchen." 1988. *Grand Rapids Press*, November 23, 4–4, 34–38.

Gofton, Leslie. 1986. "The Rules of the Table: Sociological Factors Influencing Food Choice." In *The Food Consumer*, ed. C. Tiorson, L. Gofton, and J. McKensie, 33–52. New York: Wiley.

Goodman, Ellen. 1984. "Turkey Is Just an Extra: We Feast on Family." *Grand Rapids Press*, November 28, C1.

Haley, Alex. 1976. *Roots*. New York: Doubleday.

Harger, Jim. 1991. "Soft Heart, Hard Head Makes Him Dedicated Figure in Gleaning Field." *Grand Rapids Press*, March 10, C3.

Harris, Marvin. 1985. *Good to Eat. New York: Simon and Schuster.*

Hertzler, Ann. 1988. "Gatekeeper Channel Theory: Control vs. Support." Paper presented at the meeting of the Association for the Study of Food and Society. College Station, Texas, May.

Hoogterp, Bill, Jr., and Jason Lejonvarm. 1990. *Hunger and Homelessness Action*. Minnesota: Campus Outreach Opportunity League.

"Hunger Numbers." 1989. *Why*, Spring, 14–15.

Kahn, Miriam. 1986. *Always Hungry, Never Greedy*. New York: Cambridge University Press.

Kalcik, Susan. 1984. "Ethnic Foodways in America: Symbol and Performance of Identity." In *Ethnic and Regional Foodways in the United States*, ed. Linda Keller Brown and Kay Mussell, 37–65. Knoxville: University of Tennessee Press.

Kellog, Sarah. 1991. "Recession, Welfare Cuts Mean Many More Empty Plates at Thanksgiving." *Grand Rapids Press*, November 25, B3.

Kittler, Pamela, and Kathryn Sucher. 1989. *Food and Culture in America*. New York: Van Nostrand Reinhold.

Lowenberg, Miriam E. 1974. *Food and Man*. 2nd Ed. New York: John Wiley.

Lynd, Robert S., and Helen Merrell Lynd. 1929. *Middletown*. New York: Harcourt Brace and Company.

———. 1937. *Middletown in Transition*. New York: Harcourt Brace and Company.

McCarthy, Coleman. 1990. "Imperfect Nation Lets 'Imperfect' Food Go To Waste." *Grand Rapids Press*, November 2, A7.

McIntosh, Alex. 1992. Personal communication.

McIntosh, William Alex, and Mary Zay. 1989. "Woman as Gatekeeper of Food Consumption: A Sociological Critique." *Food and Foodways* *3*(4): 317–332.

Mennell, Stephen. 1985. *All Manners of Food*. New York: Basil Blackwell.

Mintz, Sidney W. 1985. *Sweetness and Power*. New York: Viking Penguin.

Moore, Harriet Bruce. 1957. "The Meaning of Food." *American Journal of Clinical Nutrition 5*:78.

Morris, Bob, and Michael Rosenblum. *Culture and Cuisine*. New York: WNET Television, Films for the Humanities. Videotape.

Murcott, Anne. 1988. "Sociological and Social Anthropological lApproaches to Food and Eating." *World Review of Nutrition and Diet 55*:1–40.

"The Numbers: Hunger and Poverty." 1988. *Food Monitor*, 31.

Orwell, George. 1937. *The Road to Wigan Pier*. Hamondsworth: Penguin.

Pelto, Perti J., and Gretel Pelto. 1986. "Culture, Nutrition and Health." In *The Anthropology of Medicine*, ed. L. Ross, D.E. Moerman, and D. Tanard, 173–200. New York: Praeger.

Poopendieck, Janet. 1986. *Breadlines Kneedeep in Wheat*. New Brunswick: Rutgers University Press.

Sanjur, Diva. 1982. *Social and Cultural Perspectives in Nutrition*. Englewood Cliffs, N.J.: Prentice Hall.

Seymour, Diane. 1983. "The Social Functions of the Meal." *International Journal of Hospitality Management* 2(1):3–7.

Shackelford, Arn. 1992. "Food Banks Stretch Dollars, Help More." *Grand Rapids Press*, January 23, E1.

Skylar, Holly. 1991. "The Truly Greedy III." *Z Magazine*, June, 10–12.

Spence, J. 1977. "Ching." In *Food in Chinese Culture*, ed. K.C. Chang. New York: n.p.

Tannahill, Reay. 1973. *Food in History*. New York: Stein and Day.

Vitale, Alex S. 1990. "Food Not Bombs." *Z Magazine*, December, 21–22.

Walker, C., and G. Cannon. 1984. *The Food Scandal*. London: Century.

Wilber, Vincent. 1984. "Food for Thought." *Food Monitor*, Winter, 51.

Notes

[1] 1.By the sixteenth century, with the colonization of the New World, sugar was becoming part of the triangular trade and a main mechanism of generating the profit or primary accumulation which formed the financial basis for much of the success of the industrial revolution. Mintz writes: "The decline in the symbolic importance of sugar has kept almost perfect step with the increase in its economic and dietary importance. As sugar became cheaper and more plentiful, its potency as a symbol of power declined while its potency as a source of profit gradually increased" (1985:95).

[2] 2.These are not reform efforts. They are merely short-term [ones] that leave the real problem unsolved and give positive P.R. to corporate kindness.

Chapter 8

Agriculture Technology: High/low, Profits and People

> For all our technological wizardry, we human beings still owe our existence to a few inches of topsoil, an occasional thunderstorm, and a handful of crops.
>
> —Carey Fowler and Pat Mooney

Implied in the chapter title is the possibility of a conflict between the technologies of oligopolistic capitalism and the long-term needs of the general populace. Although these are not always dichotomous, this chapter generally argues that the direction oligopolistic capitalist agriculture has chosen is usually not in the long-term interests of the planet's many populations or the environment.

The forms of ownership and technology are relevant. In this chapter, oligopolistic ownership relations couple with the high technology of advanced industrial society. Ultimately we argue that the desire for short-term profit (in oligopolies) leads to technologies that conflict with the health needs of the general population and the planet. This sets the context for the examination of what economists term "externalities"—the disadvantages of this agroeconomy.

As a counterbalance to the heavy-profit/high-technology orientation, we propose some "people's technologies." China has led the way in large-scale food production. And there are several experiments around the world in low-technology, low-energy input farming. It is possible that many of these may hold the technological keys to an alternative farming future.

A Summary of the History of Agricultural Technology

In the earliest known form of society, hunting and (12000 B.C. to A.D. 7000), food getting was basically gathering. Although anthropological male bias has glorified the role of the hunter, female "gathering was clearly more important" (Fowler and Mooney 1990, 7). In the arid tropics, about 70 percent of people's caloric requirements came from gathering.

Ninety percent of humankind lived in hunting and gathering societies where they had easily available, abundant food. The !Kung Bushmen (a hunting and gathering tribe of southern Africa) spend approximately eighteen hours per week getting food. Of all subsequent societal forms, hunting and gathering was the most leisured.

Later, food growing evolved from small gardens with human labor power (horticultural society, 7000 B.C. to 3000 B.C.) to larger fields with some animal power. This agrarian society (3000 B.C. to 1850 A.D.) lasted until the introduction of industrial technology into first the factory and later the field.

In the preagricultural (and pre"scientific") context, many advances were made by practitioners. Harlan (in Kloppenburg 1990, 185) credits the American Indian with a "magnificent performance in the improvement of maize, potato, manioc, sweet potato, peanut, and the common bean." Genetic manipulation is not the sole product of contemporary "scientific" society.

While we now obtain 75 percent of our cereal from wheat, rice and maize, "prehistoric peoples found food in over 1,500 species of wild plants, and at least 500 major vegetables were used" (Mooney 1980, 304). The genetic diversity of our food sources has consistently narrowed.

Family-Farm Technology

It has been traditional to sing the praises of America's small farms. In opposition to bigness, Harwood (1992) reports

Abraham Lincoln as saying: "The ambition for broad acres leads to poor farming, even with men of energy. I scarcely ever knew a mamouth farm to sustain itself, much less to return a profit. . . . I have more than once known a man to spend a respected fortune upon one; fail and leave it and then some man of modest aims, get a small fraction of the ground, and make a good living upon it."

Family farming bred both a family/community culture and an ecological, interdependent care of the land. As Wendell Berry writes: "The culture that sustains agriculture and that sustains it most forms its consciousness and its aspiration upon the correct metaphor of the 'Wheel of Life'" (Berry 1986, 89). Family farming conformed to certain basic ratios between "eyes and acres" (Berry 1986, 89). Berry argues that "land that is in human use must be lovingly used." Family farming utilized conserving technologies of intercropping, recycling manure, and genetic selection of appropriate crops.

The family farmers who worked land over generations knew its history. Unlike the segmented industrial labor of "dismembered gestures," farming involved the totalistic approach of the artist. Family farming is "one of the last places where the maker is responsible from start to finish for the thing made" (Berry 1986).

Berry attributed the decline of the family farm to a decline in family values. He sees the problem in farmers' acceptance of the assumptions that capitalist values are the only motivators. He would prefer that the federal government actively support the family farm. Unfortunately, the truth is usually just the opposite.

The Technology of Oligopoly

For a long time it was assumed that the peculiar nature of farming inhibited the penetration of capitalist relations of production into the countryside (see Mann 1990).

While industrial enterprise evolved from competitive to oligopoly capitalist forms of ownership, much of the farming sector of the economy remained among small entrepreneurs.

Agriculture in the Industrial Era

It was through seed (Kloppenburg 1990) that industry entered the agricultural area. Pioneered by private agricultural extension agents, the U.S. Department of Agriculture began systematic application of science to seed selection: "Hybridization came increasingly to be used in conjunction with selection" (Kloppenburg 1990, 78).

After the Great Depression, the Adams and Purnell acts ensured the continuation of high-level scientific involvement in agricultural research on increasing yields. As Kloppenburg writes: "The development of hybrid corn has long been regarded as the supreme achievement of public agricultural science" (1990, 91). Thus, "by 1965 over 95% of U.S, corn acreage was planted with the new seed."

Seed hybridization solved the problem of capitalist technology's penetration of the agricultural sector. Whereas, before hybridization, farmers reused their seed and adjusted their choices on the basis of soil and weather conditions, hybrid seed was controlled by the seed companies. Because profitability is key in a capitalist system, farmers simply could not refuse the more productive hybrids. Refusal meant being outcompeted and elimination from the business. Oligopolistic capitalism had finally brought agriculture within its realm of operation.

These developments gave new life to moribund seed companies. As they grew in power, they chose among the scientific developments of public research. And they began to conduct their own research. Even more important, hybrid seed became the vehicle for controlling the application of other agricultural inputs. Presaging Green Revolution technology, certain hybrid seeds presupposed a management system that required "mechanization and the application of agrichemicals" (Kloppenburg 1990, 117). Once again, capitalist industrial technology had found a way to dominate a previously craft industry as it fostered allies in the petrochemical and agricultural machine industries.

The Green Revolution portrayed itself as an attempt to end the world's food problem. The assumption was a shortage of food in the developing world. If Western science could "reform" the food production process to increase yields, world hunger would be solved. If, in the process, the new corporate oligopolies enriched themselves, so much the better for corporate capitalism.

The Green Revolution began with Rockefeller Foundation sponsorship of the CIMMYT wheat research facility in 1943 in Mexico. In 1956 the Ford Family Foundation sponsored a program in India. In the early 1960s both foundations combined to sponsor IRRI (the International Rice Research Institute) in the Philippines (Mooney 1980, 39). Kellogg joined in transferring sponsorship to the World Bank and an amalgam of UN agencies and foundations, resulting in the formation of CGIAR (Mooney 1980, 39). As Mooney writes: "The foundations successfully shifted the financial burden, while maintaining considerable influence within the CGIAR. For example, all but one of the international crop research station directors have come up through the 'foundations'" (1980, 39).

With government (the state) performing its accumulation and legitimation functions simultaneously (O'Connor 1973), the individual state and the world government organizations both facilitated capital growth and the (perhaps illusory) position of helping the developing world.

The accumulation function happened through the development of specific types of seeds. These seeds, in addition to requiring replacement (purchase) every year (as hybrids), also required large inputs of petrochemicals (fertilizers, herbicides and fungicides). Corporations producing these other inputs blossomed.

Again, because under the conditions of ideal inputs Green Revolution seeds were so much more productive, those large farmers who could afford all the accompanying chemicals and irrigation became prosperous. And those who could not (usually small farmers) lost the competitive race. As a result, the petrochemical industry grew in influence throughout world agriculture. Fertilizers, pesticides, and herbicides became a worldwide phenomenon. And richer farmers prospered to the detriment of the small and poorer ones.

In addition, U.S. manufacturers experienced "increased demand for U.S. farm tools ...irrigation pumps and other agricultural equipment" (Mooney 1980, 41). As Mooney writes:

> By the sixties, agricultural enterprises were in need of a new market to maintain their growth. Bilateral and multilateral aid programs made expansion into the Third World financially possible. Twenty years later, major agrichemical firms have achieved a worldwide distribution system able to market successfully in Asia, Africa and Latin America. The Green Revolution was the vehicle that made all this possible. (1980, 41)

The methods and techniques (technologies) of farming were now set, not by the wisdom of individual farmers, but "scientifically" by the oligopolies that have come to dominate agriculture. While at first true only in the United States, oligopolization and vertical and horizontal integration have now spread throughout the world. This situation means increased profits for multinational producers of agricultural inputs and decreased work lives for small producers who cannot compete with this technology.

Giesler and Popper note that "In 1978, the largest 5% of land owners owned 75% of the land" (1984, 8) and Manchester writes that "overall the percent of agricultural output vertically integrated or coordinated has risen from 19% in 1960 to a little over 30% in 1980" (1983, 105). After years of resisting capitalist forms of the industrial organization of food production (see chicken production in chapter 1), American food production is now in the process of a fundamental conversion.

These trends continue with the arrival of the newest and potentially most profitable development, biotechnology. As Kloppenburg writes:

> New . . . techniques such as protoplast fusion and recombinant DNA transfer allow direct access to a discrete piece of a plant's genome at the cellular and even the molecular level. It is becoming possible to change gene frequencies with a wholly unprecedented specificity, and such recombinations are no longer limited to organisms that are sexually compatible. New plant varieties are being engineered in the strongest sense of the word connotations of precision and foresight. (1990, 192).

The potential significance of these developments is so great that the U.S. Office of Technology Assessment maintains that the United States is "at the brink of a new scientific revolution that could change the lives and futures of its citizens as dramatically as did the Industrial Revolution . . . and the computer revolution" (Busch, Burkhardt, and Lacey 1991, 3).

Initially, corporate users of this technology will probably turn to increased hybridization of previously unhybridizable crops. Because farmers still plant a considerable proportion of their crops from the previous year's seeds, these areas represent a significant potential for profitable takeover.[1]

In addition, research has focused on the control of weeds. Basically, biotechnology researchers have sought to make seeds resistant to petrochemical herbicides so that only the weeds would be affected—and more petrochemicals can be used. Other work (of less intensity) is being done in attempting to engineer resistance to insects and disease. Some success in resisting fungal diseases has been reported (Busch and Lacy 1991, 9).

In terms of the prospective dangers to corporate capital, there exist also the possibilities of destroying the chemical industry and of taking production out of the field and into the laboratory. In one example, "the development of a thaumatin product through *genetic engineering* may continue a transition to alternative sweeteners eliminating the market for beet and *cane sugar* and capturing the valuable sweetener market, currently worth $8 billion in the U.S." (Busch and Lacey 1991, 10).

Once again, a realignment of the respective roles of government and private industry is happening. Because biotechnology has now become highly profitable, capitalist industry has taken to maintaining its own research institutions and relegating (when possible) to public universities the less profitable projects.

In addition—again simultaneously performing accumulation and legitimation functions—the U.S. Congress passed, in 1970, the Plant Variety and Protection Act (PVPA), which protects the private ownership of novel genetic forms. Some patents have already been approved. The line between pure and applied research is once again becoming blurred. As Busch and Lacey write: "These biotechnology arrangements—between firms, universities, faculty members, and student/trainees—included large grants and contracts between companies and universities in exchange for patent rights and exclusive licenses to discoveries" (1991, 14). This situation raises serious questions about the role of basic and applied research. Who sets the agenda? In whose interest(s) is research done? What if discoveries (such as herbicide-free weed resistance) appear? Will the research become public?

Like hybridization and the Green Revolution, biotechnology raises questions about interests, externalities, and potential problems of the technology. It is to their examination that we now turn.

An Appraisal of "Scientific Agriculture"

Jeffersonian America was a nation of small farmers. Involved were highly skilled artisans of agriculture who utilized traditional and practical wisdom in producing the basic foods for the country. But as industrial practices penetrated farming, small farms succumbed to competition from bigger and bigger ones: "From 1945 to 1973 [farms] decreased by 55% from 5.9 million to 2.8 million. . . . By 1981 there were 25,000 superfarms, representing 1% of all farms and receiving 66.3% of total net farm income" (Busch and Lacey 1991, 23).

Farming and the industries associated with it have consistently moved from small town, small farm to large oligopolies. As Busch and Lacey write: "Overall, in 1979, 15 companies accounted for 60% of all farm inputs, 49 companies processed 68% of all the food, and 44 companies received 77% of all wholesale and retail food revenues" (1991, 23).

This process of oligopolization and fewer farms combined with the Green Revolution and biotechnology to form significantly different farming patterns from those of early America. Before industrial farming, American farmers utilized intercropping (planting a number of crops in the same field), organic fertilizers (manure and compost), the planting of multiple varieties of the same crop, the rotation of crops (taking and contributing different nutrients to the soil) and planting multiple varieties (e.g., five kinds of corn). Animals (horses and oxen) were the basic power. They did not pollute and used no petrochemicals.

The industrialization of agriculture has replaced these methods with monocropping (planting the entire field with the same variety of one, high-energy crop). This process uses energy with its farm machinery (both the energy involved in building it and in running it), heavy petrochemical application (often sprayed by airplanes), and the elimination (insofar as possible) of anomalies in the land (trees, hills, waterholes) so as to make possible central pivot irrigation and uniform plowing and mechanical harvesting.

The result is pollution of water tables, soil depletion, and genetic vulnerability. Wes Jackson writes: "An increasing number of people are finally recognizing that agriculture itself is an ecological problem outranking industrial pollution" (1980, 91). And, as *Alternative Agriculture* states, "the Environmental Protection Agency has identified agriculture as the largest nonpoint source of water pollution. Pesticides and nitrates from fertilizers and manures have been found in the groundwater of most states" (Committee on the Role of Alternative Farming 1989, v).

Unfortunately, the high-response seeds of the Green Revolution and biotechnology generally require more fertilizers than the original seeds (Fowler 1990, 130). As a result, Jackson reports "14,000 individuals are non-fatally poisoned by pesticides each year and 6,000 are injured enough to be hospitalized" (1980, 27).

Because many of the scientifically developed seeds presupposed very specific inputs, increased irrigation has become problematic. In California there is a water shortage. Irrigated soils often demonstrate problems of mineralization and salinization. Ultimately less land will be available for agriculture. In addition, biological reactions to chemicals are often adaptive. Kloppenburg quotes Sommers: "Some forty-one weed species now show resistance to herbicides" (Kloppenburg 1990, 247). And nearly three quarters of chemically induced crop diseases have been caused by herbicides.

Soil depletion is another disbenefit. Preindustrial farming husbanded its soil through crop rotations, organic fertilizers, and natural curvatures of the land (hills, trees, irregularities). Jackson cites an Iowa State study: "in thirty-eight years the gross per-year soil loss increased from 3 to 4 billion tons per year but the *average* per acre loss jumped from 8 to 12 tons" (1980, 17). This chemical dependent form of agriculture depletes the environment in ways that mortgage the future of American agriculture.

Of even greater import is the issue of the reductions of the multitudinous varieties of most crops and genetic vulnerability. Previous forms of agriculture had maintained a large variety of the same crops. For instance, Andean farmers cultivate some 3,000 of the 5,000 known varieties of potatoes. "Sometime 45 distinct varieties can be seen growing in a single field" according to Fowler and Mooney (1990, 19).

Multicropping varieties makes possible defenses against pests, weather, and other assaults from nature. Extreme dryness will be good for some varieties. Those that require more moisture will not do well that year. But in another year it may be wet, and the reverse will be true. The result is a more or less constant production of a crop.

When only one or a few varieties are grown, the crop is much more vulnerable. This happened in the famous Irish potato famine. When *phytophtora infestans* (potato blight) infected the Irish crop, the entire crop was affected. Because the whole crop descended from only two varieties of potatoes (imported from the Americas), there was no repertoire of genetic resistance. Only the importation of different genetic varieties of potatoes saved potatoes as a basic crop for Ireland. As Fowler and Mooney write: "Potatoes were the first crop in modern history to be devastated by lack of resistance . . . and the first crop to be rescued by the wealth of defenses built up over thousands of years of diversity" (1990, 5).

Since then epidemics have hit cotton (1917 and the 1980s) and the Indian rice crop (1943) and left the U.S. cultures of oats 80 percent depleted in the 1940s and 1950s (Fowler and Mooney 1990, 47). In 1970, corn blight threatened the U.S.'s grain deal with the USSR. As Fowler and Mooney state: "Each time resistance was needed. And each time it was found in the centers of diversity in land races [peasant-grown varieties] that had somehow escaped homogenization, or in those crops' wild relations" (1990, 47).

Many scientists have felt a need to provide for storage of many of these genetic resources. Seed banks exist throughout the world that periodically grow original seeds. But there have been many instances, often due to human error and underfunding, of electrical and refrigeration failures, resulting in loss of these seeds. There at present exists a National Seed Storage Laboratory at Fort Collins. It is essentially a "genetic Fort Knox" (Fowler and Mooney 1990, 187). It makes its resources available to the oligopolies, which proceed to make profits from them.

An additional problem is the exploitation by developed world private oligopolies of the genetic resources of the developing world. Virtually all the centers of diversity in the world (Vavilov centers) are found outside the developed world. Yet profit-oriented researchers assume it is their right to take these genetic resources back with them for private manipulation and profit.[2]

Of greater contemporary importance is the manner in which biotechnology allows plant scientists to select genetic material from the rest of the world to protect or solve certain problems in present crops. These genetic pieces are often worth large financial amounts to the companies doing the genetic engineering. Kloppenburg reports several examples:

> A Turkish land race of wheat supplied American varieties with genes for resistance to stripe rust, a contribution estimated to have been worth $50 million per year (Myers 1967, 68). The Indian selection that provided sorghum with resistance to

> greenbug has resulted in $12 million in yearly benefits to American agriculture. An Ethiopian gene protects the American barley crop from yellow dwarf disease to the amount of $150 million per annum (*New Scientist* 1983, 218). . . . And new soybean varieties developed by University of Illinois plant breeders using germplasm from Korea may save American agriculture an estimated $100 - $500 million in yearly processing costs (*Diversity* 1986b, 218). . . . It is no exaggeration to say that the plant genetic resources received as free goods from the Third World have been worth untold *billions* of dollars to advanced capitalist nations (1990, 167-169).

If such profits are to be made from the genetic resources of the developing world, should there not be some repayment?

Furthermore, the possibility exists of ruining the cash crop of some products of the developing world. We have already mentioned the problems that could be caused for the Caribbean if most sweeteners become artificial ones. Again, as Kloppenburg writes:

> Industrial plant tissue culture offers developed nations the possibility of replacing imports of tropical plant products with domestic production. It permits effective appropriation of a new class of Third World germplasm, that of plants that will not grow in the temperate climates of the Northern Hemisphere but whose cells can be grown in stainless-steel fermentation vats anywhere (1990, 274).

If this practice were to become widespread for the large number of scents, spices, dyes, and drugs that the developing world supplies, what would be the effect on these already weak and unstable economies?

In sum, the Green Revolution and biotechnology have been highly profitable for the oligopolies that dominate the economy of the developed world. But their significant side effects may mortgage our planet's future. They pollute. They make profits from depleting our genetic heritage. And they offer the potential destruction of much of the agriculture of the developing world.

While these scientific technologies have been touted as solutions to the food problems of the developing world, they have in fact really benefited the investors in U.S. and other developing world oligopolies. They have made the developing world dependent on expensive inputs of petrochemicals and seeds produced by private oligopolies. And they have stimulated the same process of destruction of small subsistence farmers to the benefit of large, corporate-sponsored farmers. The U.S. agricultural pattern has become increasingly worldwide.

It is really to "blame the victim" when one remarks that small farms are "failing" around the world. The more accurate analysis is the Lappe and Collins paradigm (see chapter 9) where large farmers, often sponsored by foreign interests, push "uncompetitive" small farmers off their land and into the substratum of incomeless homelessness. This is the fundamental fact of agriculture in the capitalist world.

Agricultural Ideas for People and the Planet

Potential solutions for both developing and developed worlds come from basically two sources. The first are the agricultural techniques developed in parts of what is left of the socialist world that have succeeded in feeding all their people. The second set of solutions come from organic and experimental farmers in the developed world. Many of these experiments are already producing successful results.

Because of the scale of its success, especially by comparison with capitalist India, China provides one example of a people's ecological agriculture. Other socialist examples (Cuba, Nicaragua and Kerala) appear in chapter 10 as success stories with food.

By way of background, one should know that China did not have the options of the developed world when it went about formulating an agricultural policy after its 1949 revolution. Even if it had wanted to, it could not have imitated the Green Revolution's high petrochemical input strategy of India and the developed world. Instead, China decided to build "socialism in one country" without getting into relations of dependence. Therefore it made significantly different choices in the areas of energy, recycling, crop choice, and social technology (the labor organization of the farms).

Though much of China's arsenal seems novel (by world capitalist standards), a lot is simply the employment and enhancement of traditional farming techniques supplemented by what contemporary scientific knowledge was relevant to them. For instance, an agricultural manual for the first century B.C. lists the following agricultural practices:

- Multiple cropping: i.e., winter wheat followed by summer millet
- Seed pre-treatment: soaking seeds in organic fertilizers and herbicides
- Adjusting the correct temperature of water routed through rice paddies
- Recycling into fertilizer anything with nitrogen content
- Evaluation of appropriate soils for specific crops (Anderson 1988, 49)

The large varieties of rice in China necessitated labor-intensive agriculture. Similarly, because there was a large population, China decided to do labor-intensive rather than heavy-machinery, capital-intensive agriculture.

Best known in the world is the small (28- or 35- horsepower) tractor that looks a bit like a big lawnmower. It can plow on hills and can pull a trailer on roads. Its design comes from people and engineers who first looked at the real needs of farmers before designing the technology.

China also continues to utilize traditional nonpetrochemical sources of power for agricultural locomotion. Oxen still plow the fields. And the water buffalo is a source of labor in rice paddies. In addition, pest control is labor-intensive rather than predominantly petrochemical. As Lappe and Collins write, in China, "pests are controlled before they become a serious problem" (1978, 73). Aided by skilled agronomists, youth teams become an early warning system for crop damage. They have brought locust invasions under control and reduced to 1 percent the damage from wheat rust and rice borer (1978, 73). This is in contrast to massive pesticide and herbicide aerial spraying in Western forms of agriculture.

Fertilizer usage in China has moved toward petrochemicals because of the present availability of petroleum. Nevertheless, the Chinese still do a major job of recycling "green manure" (human feces) into fertilizer. The countryside adjacent to most cities has an ample supply of this fertilizer transported out to it.

Similar attention has been paid to the sources of energy for locomotion and cooking. Bio-gas (largely methane) is produced in 4 million bio-gas pits and "used for cooking, lighting, and running farm machinery" (Lappe and Collins 1978, 179). This cheap and abundant energy comes from animal dung. Although China has moved to create more petrochemical fertilizers, it has used them on a self-sufficient basis, rather than import them as does India.

Traditionally the Chinese diet has been grain based: "Food grains have consistently supplied some 86%–89% of available protein" (Riskin 1990, 331). But, unlike countries relying on grains spawned by the Green Revolution, China has chosen traditional grains. "Rice, maize, sorghum, and millet produce far more under most conditions than do wheat and barley" (Anderson 1988, 128). There was a focus on soybeans, a traditional Asian staple and a high protein source. Unlike the beef cultivators of the West, what protein the Chinese got from animal flesh came from "pigs and chickens—excellent converters of cheap, inferior food into meat" (Anderson 1988, 129).

Coequal with technology has been the considerably more egalitarian ownership of land and power among the Chinese people. Because land reform eliminated the rich peasants and landlords, most people have an interest in producing. Their lot will be better for it. In contrast, even such an appropriate technology as bio-gas works to the detriment of the poorest in capitalist India. When dung has financial value, it is taken by the richer Indian peasants for bio-gas plants, leaving the poor with no source of fuel at all (Lappe and Collins 1978, 178).

China has come out much better. As a result of its agricultural practices, "China feeds 50% more people 20% better with 30% less cultivated land" (Lappe and Collins 1978, 198). The most recent moves in the direction of capitalist economic organization are grafted on an agricultural system that is labor-intensive, that is based on recycling and low technology, and that operates generally for the benefit of everyone rather than a small minority of oligopolists who profit from the Green Revolution and biotechnology in the oligopolistic capitalist world.[3]

Promising U.S. Experiments and Projects

Two prominent spokespeople in the United States for alternative agriculture are Wendell Berry and Wes Jackson. The Rodale Organization is also conducting a number of interesting experiments. And other experimental farms around the country show promise.

Wendell Berry, an English professor at the University of Kentucky, farms in a spiritual and "orthodox" (for previous generations) way. Geneticist Wes Jackson's Land Institute in central Kansas conducts experiments in perennial grasses as alternatives to the high-technology, high-chemical agriculture that characterizes the American mainstream of big farmers.

Both have contributed suggestions for technological changes. Both suggest some socioeconomic adjustments to allow the alternative technology to happen. And both conceive of agriculture as not divorced from a fundamental spirituality.

Out of a concern for the massive soil erosion that contemporary large agriculture generates, Jackson has been experimenting with no-till agriculture. The vehicle has been the intercropping of perennials. For this he has chosen eastern gamma grass and Maximilian sunflowers. The former is full of protein and amino acids. Jackson speculates that "its seed yield can now be increased from fifteen to thirty fold, because some of the energy in the leaves can be reallocated into the seeds" (in Schneider et al. 1986, 88). The sunflowers provide a natural herbicide. Together they may make an ecologically balanced combination that will not have to be planted every year. Jackson claims he is imitating nature's own solution to erosion and food production.

Wendell Berry is an admirer of the Amish. From them he takes certain spiritual values combined with appropriate technology. One answer to the massive use of petrochemical energy sources on farms is a return to horsepower. As Berry writes: "There are certain problems for which the use of horses is the appropriate solution—or for which we have so far found no more appropriate solution" (1977, 203).

Furthermore, Berry believes in a synthesis between the older methods of farming and new techniques discovered by contemporary organic farmers. Berry speaks in favor of "the possibility of grafting the soil management methods of the more advanced organic farmers upon the traditional structures and skills of the old horse-powered farming" (1977, 210).

The most prominent institutional experimentation on organic or regenerative farming methods is now done by the Rodale Research Center in Maxatawny, Pennsylvania. For instance, "researchers have discovered how mulching around cabbages can cut down on flea beetles and how hardy, fast-growing varieties of Oriental vegetables can provide a profitable second crop in the early fall. Planting red clover, rich in nitrogen, between corn rows, they have found, reduces both erosion and the need for chemical fertilizers" (Schneider et al. 1986, 51). In a very real sense, they are doing the research that a government which governed in the interests of small farmers should be doing. They have generated some governmental interest and research. Agricultural colleges in Oregon, Nebraska, Iowa, North Carolina and California are working on "reducing the nation's $10 billion annual farm chemical bill, improving soil nutrition and increasing profits" (Schneider et al., 1986, 51).

In addition, there are a number of other experimental and profitable farming experiments around the country. One such example is the Harder Farm near Greatly, California. Here Dick Harder practices regenerative no-till agriculture with rice and with kiwi fruit. Although his yields are a bit lower than commercial agriculture, the savings in the cost of inputs make him competitive. As he says, "'I like to think of the long-term advantages of this kind of farming. . . . I am building my soils every year instead of taking from them'" (in Schneider et al. 1986, 55–56).

Another successful example is the Gasconade Farm near Vienna, Missouri.

> Rather than relying on one crop that is harvested all at once, the Gasconade operation challenges the assumptions of traditional breadbasket agriculture by producing a variety of vegetables, herbs, flow-

> ers and fruits that can be harvested continuously and on a large scale . . . . Last year [1985] Gasconade sold fifty-two different types of vegetables, herbs and flowers to six St. Louis restaurants, including everything from fancy lettuce to cherry tomatoes, spinach, golden peppers and squash. (Schneider et al. 1986, 86).

These experiments demonstrate the feasibility of regenerative/organic agriculture. The problem is that the oligopolies influencing government research directions have not provided serious support for this method of farming.

There are some private attempts to support organic/regenerative farming. Twenty miles north of Albany, New York, Janet Britt has pioneered an ownership/workshop pattern called Community Supported Agriculture. About 100 shareholders pay in advance for the vegetables she will deliver to them. If natural disaster strikes, all share the loss. But under normal circumstances, all get less expensive vegetables grown without petrochemicals. There are now community-supported agricultural organizations in California, Massachusetts, Michigan, New Hampshire, New Jersey, New York, Pennsylvania, Vermont and Wisconsin (Cook 1990, 57).

Wes Jackson has proposed a people's corporate mode of ownership: land trusts. "As small farmers are forced out, the trusts can buy the land and give priority to those who have owned the land, asking them to stay and even offering them opportunity to buy into the trust. The task of the owners would be to draw up the rules on how the farm is to be managed, to ensure the bottom line is not profit but healthful conservation" (Jackson 1980, 90–91).

For Wes Jackson and Wendell Berry, the issue is both material (technological) and spiritual. Jackson notes the fundamental antagonism between humans and nature in the Judeo-Christian tradition. Taoism, Buddhism, and Jainism are far more sympathetic to the natural world as part of the religious experience (1980, 66).

Berry notes our history of exploiting nature rather than of *mutual nurture*.[4] And he is stridently critical of some contemporary uses of food. "Food is not a weapon. To use it as such—to foster a mentality willing to use it as such—is to prepare, in the human character and community, the destruction of the sources of food. The first casualties of the exploitive revolution are character and community" (1977, 9).

Berry is concerned that a way of life is being destroyed by the contemporary forms of agriculture: "The word *agriculture*, after all, does not mean 'agriscience,' much less 'agribusiness.' And *cultivation* is at the root of the sense of both *culture* and *cult*. The ideas of tillage and worship are thus joined in *culture*" (Berry 1977, 87).

In a very real sense, the concomitant problem of big agriculture is the elimination of a way of being in the world that, from the Jefferson era, often characterized America. Big farming is destroying a way of life rooted in care for the soil and for nature. Instead we have come to accept as normal a fundamentally unecological, exploitative relationship to nature that degrades both us and nature. We need "'symphonic' agricultural systems . . . [which involve] concurrent attention to ecological sustainability, economic viability, and equity" (Harwood 1992, 24).

Only a multilevel, pluralistic approach will begin to remedy the multiple maladies of industrial capitalist agriculture.

Summary

Idealist social science might expect the weight of Berry's and Jackson's arguments (ideas) to affect change in the American corporate farm economy, but the material fact remains that big agriculture is at present immensely profitable to the capitalist oligopolies. The process of "mining the fields" rather than nourishing them may be shortsighted. But it now supplies large profit margins in the corporate economy.

There is, it is true, a degree of pluralism in eating. Such experiments as the Gasconade Farm cater to the epicures and health food constituencies in America. And, in a small way, they have provided the demand for better nutritional products and better agricultural products. There has even been some movement in the direction of organic products (See chapter 10).

To the "uninterested" (in an economic sense) academic, technological aid to the developing world might better be abandoned by governments and oligopolies. It seems that the choice of technologies to export always involves the most profitable for the oligopolist that control seed, petrochemical, farm-tool manufacturing.

Instead, one might hope that impartial academics might attempt to synthesize some of the insights of biotechnology with the indigenous agricultural processes and products of many developing countries. One must explore Frances Moore Lappe's insight that almost every country in the world has the resources to feed itself—at least on a basically vegetarian diet (see chapter 9).

Many relatively small organizations—such as Oxfam International—have built viable reputations providing many practical, low-cost and low-technology solutions to pressing problems. They should be encouraged but not controlled.

China has provided the most impressive model for the developing world. It essentially dismissed outside help and developed a combination of traditional agricultural practices and scientifically engineered adaptations to its own native agricultural products and techniques. And, unlike India, which relies heavily on foreign technology and aid, China has successfully fed its entire population (except for the period of its Cultural Revolution) for a couple of decades now.

Small and developing countries should not be forced to reject the enormous possibilities of biotechnology. Rather than have the direction of its research controlled by the profit-driven oligopolies (often in combination with the land grant agricultural farm institutions), biotechnology research centers controlled by various people's elements in the developing world should be established. China might provide some biotechnological solutions in the interests of the whole people.

While one might wish for a change of heart (ideas) among those few multinational giants who provide or influence the basic direction of most developed countries' farm policies, the most we can probably hope for is some degree of pluralism among the diaspora of people who care about nature and the earth.

References

Anderson, E.N. 1988. *The Food in China.* New Haven: Yale University Press.

Belden, Joseph N. 1986. *Dirt Rich, Dirt Poor: America's Food and Farm Crisis.* New York: Routledge and Kegan Paul.

Berry Wendell. 1977. *The Unsettling of America: Culture and Agriculture.* San Francisco: Sierra Club Books.

———. 1986. "In Defense of the Family Farm." Speech at Kalamazoo College, Kalamazoo, Mich., October 8.

Bourlag, Norman. 1978. "Genetic Improvement of Crop Foods." In *The Feeding Web*, ed. Joan Dye Gussow, 395–399. Palo Alto, Calif.: Bull Publishing Company.

Buchanan, Anne. 1982. *Food, Poverty and Power.* Nottingham: Spokesman.

Busch, Lawrence, William B. Lacey, Jeffrey Burkhardt, and Laura Lacey. 1991. *Plants, Power and Profit.* Cambridge, Mass.: Basil Blackwell.

Committee on the Role of Alternative Farming Methods in Modern Productive Agriculture. 1986. *Alternative Agriculture.* Washington, D.C.: National Academy Press.

Cook, Jack. 1990. "Consumers are Getting Healthy Produce Direct from the Field by Becoming Partners with the Farmers Who Feed Them." *Harrowsmith Country Life*, May/June, 53–57.

DeLind, Laura. 1990. "Agriculture, Education and Industry: A Progressive or Problematic Partnership for Michigan?" *Michigan Sociological Review* 4:20–32.

———. 1992. "The State, Hog Hotels, and the 'Right to Farm': A Curious Relationship. Department of Sociology, Michigan State University.

Dre'ze, Jean, and Amartya Sen. 1989. *Hunger and Public Action*. Oxford: Clarendon Press.

Engardio, Peter. 1983. "The Practice and Promise of Organic Farming." *Food Monitor*, January/February, 19–23.

Flaherty, Diane. 1988. "The Farm Crisis: Rural America in Decline." In *The Imperiled Economy: Book II—Through the Safety Net*, ed. Robert Murphy et al., 4–17. New York: Union for Radical Political Economy.

Fowler, C., Lach Kovics, P. Mooney, and H. Shad. 1988. "The Laws of Life: Another Development and the New Biotechnology." *Development Dialogue*, 1–2.

Fowler, Carey, and Pat Moonbey. 1990. *Shattering: Food, Politics and Loss of Genetic Diversity*. Tuscon: University of Arizona Press.

Giesler, Charle G., and Frank J. Popper. 1984. *Land Reform American Style*. Totowa, N.J.: Roman and Allanheld.

Harlan, Jack R. 1975. *Crops and Men*. Madison, Wisc.: American Society of Agronomy.

Hart, Robert D., and Michael W. Sands. 1990. "Sustainable Land Use Systems Research and Development." Paper presented at Sustainable Land Use Systems Research Conference, New Delhi, India.

Harwood, Richard R. 1992. "The Structure of Biological Diversity at the Agricultural, Environmental and Social Interface: An Agricultural Perspective." Paper presented at Diversity Conference, Michigan State University, East Lansing, Mich.

Horsfall, James G. 1979. "Iatrogenic Disease: Mechanisms of Action." In *Plant Disease: An Advanced Treatise*, ed. J.G. Horsfall and E.B. Courtney, 267–294. New York: Academic Press.

Hsu, Cho-yun. 1980. *Han Agriculture*. Seattle: University of Washington Press.

Jackson, Wes. 1980. *New Roots for Agriculture*. Lincoln: University of Nebraska Press.

Kates, Robert W. 1987. *Ending Hunger for the Second Billion*. Providence: The Alan Shawn Feinstein World Hunger Program.

Kloppenburg, Jack Ralph, Jr. 1990. *First the Seed: The Political Economy of Plant Biotechnology 1492-2000*. Cambridge: Cambridge University Press.

Kutzner, Patricia L. 1990. "World Hunger: What Have We Learned?" *Hunger Notes 16*:1–2.

Lacey, William B., and Lawrence Busch. 1987. *Biotechnology and Agricultural Cooperatives: Opportunities and Challenges*. Proceeding of a workshop co-sponsored by the Committee for Agricultural Research

Policy and the College of Agriculture, U.S. Department of Agriculture, Lexington, Ky, April 5–7.

Lacey, William B., Lawrence Busch, and Laura R. Lacey. 1988. "Public Perceptions of Agricultural Biotechnology." In *Agricultural Biotechnology: Issues and Choices*, ed. Bill R. Baumgardt and Marshall Martin, 139–161. West Lafayette, Ind.: Purdue University Agricultural Experiment Station.

Lappe, Frances Moore, and Joseph Collins. 1978. *Food First*. New York: Ballantine Books.

Lincoln, Abraham. n.d. "Abraham Lincoln on Agricultural Transactions." Wisconsin State Agricultural Society, 858–859.

Manchester, Alden. 1988. *The Farm and Food Systems: Major Characteristics and Trends*. Ann Arbor: University of Michigan Extension Cooperative.

Mann, Susan Archer. 1990. *Agrarian Capitalism in Theory and Practice*. Chapel Hill: University of North Carolina Press.

Mooney, Pat Roy. 1980. *Seeds of the Earth: A Public or Private Resource*. San Francisco: Institute for Food and Development Policy.

O'Connor, James. 1973. *The Fiscal Crisis of the State*. New York: St. Martin's Press.

Office of Technology Assessment. 1987. *New Developments in Biotechnology: Public Perception of Biotechnology*. OTA-BP_BA-45. Washington, D.C.: Government Printing Office.

Otero, Gerardo. 1991. "Biotechnology and Economic Restructuring: Toward A New Technological Paradigm in Agriculture?" Paper presented at the American Sociological Association Meetings. Cincinnati, Ohio, August 23–27.

Pimental, David, and Marcia Pimental. 1979. *Food, Energy and Society*. London: Edward Arnold Publishers.

Riskin, Carl. 1990. "Food, Poverty and Development Strategies in the Peoples Republic of China." In *Hunger in History*, ed. Lucile F. Newman et al., 68–92. Cambridge, Mass.: Basil Blackwell.

Russell, Dick, and Noel Weyrich. 1986. "The Regreening of America." *New Age Journal*, March, 50–56, 86–93.

"Seed Monopoly." 1975. *Elements*, February, 6–7.

Singer, Peter. 1975. *Animal Liberation*. New York: Aron Books.

Smil, Vaclav. 1993. *China's Environmental Crisis*. Armonk, N.Y.: M.E. Sharpe.

Sommers, Charles F. 1982. "More Pests Unfazed By Chemicals." *Successful Farming*, January, 23.

Stanton, B.F. 1983. "What Forces Shape the Farm and Food System." In *The Farm and Food System in Transition*. Ann Arbor: University of Michigan Extension Cooperative Service.

Sustainable Agriculture Working Groups. n.d. "Recasting Food Aid with a Hunger Focus." Marne, New Mexico, n.p., 18.

Swaminathan, M.S. 1988. "Global Agriculture at the Crossroads." *Earth 88: Changing Geographical Perspectives*, 316–330.

USDA Agricultural Research Service. 1982. *Economic Indicators of the Farm Sector: Income and Balance Sheet Statistics* (ECIFS).

Vogeler, I. 1982. *The Myth of the Family Farm: Agribusiness Dominance of U.S. Agriculture*. Boulder, Colo.: Westview Press.

Notes

[1] 1.In the United States the figures for crops replanted from the previous year's crops are 60% of wheat, 40% of soybeans, 70% of oats, 50% of barley, and 50% of cotton (Kloppenburg 1991, 243).

[2] 2.The only crops of economic importance indigenous to the United Sates and Canada are sunflowers, blueberries, cranberries, pecans, and Jerusalem artichokes (Kloppenburg 1990, 46). Virtually all the rest of our important crops came from different parts of the world. Some of these crops and their sources include barley, coffee, and wheat from Ethiopia; asparagus, cabbage, lettuce, olives and oats from the Mediterranean; carrots, cotton, grapes, lentils, mustard, peas, and spinach from Central Asia; oranges, peaches, sugar cane and tea from China; and numerous beans, corn, tobacco, and tomatoes from Central America (Kloppenburg 1990, 48).

[3] 3.As this goes to press, there are some credible indications that Chinese agriculture may be moving in a new ecological direction. Smil (1993) reports numerous instances of soil erosion, salinization, alkalization, and dessertification (56); pollution from the increasing use of synthetic fertilizers (58); and water (58) and air (110) pollution. It remains to be seen how widespread this is and whether this will be a long-term trend.

[4] 4.One must note that family farmers were not without their own forms of exploitation of family members' free labor.

Chapter 9

World Hunger

Most people believe that there is just not enough food to go around. Yet "the world is producing, each day, two pounds of grain, or more than 3,000 calories for every man, woman and child on earth . . . 3,000 calories is about what the average American consumes. And this estimate . . does not include the many other staples such as beans, potatoes, cassava, range-fed meat, much less fresh fruits and vegetables. Thus, on a global scale, the idea that there is not enough food to go around just does not hold up"—Food and Agriculture Organization.

Forty years ago the Nazis killed six million people. At the Nuremberg trials those responsible claimed they personally could not be blamed, that it was the "system," that the decisions were made higher up. This was not accepted: they were condemned. Today it is a question of perhaps 20 million dying *every year*. These deaths from starvation are also the result of a system, of deliberate policies. What can we say of those, in government, business or international agencies, who operate this system, which results in these deaths.

—Anne Buchanan, 1982

How did the world come to tolerate hunger in the midst of plenty? How widespread is hunger? What causes it? How can it be cured? Will it be? Ever?

This chapter attempts diverse answers to these questions. It documents the extent of hunger, then divides into a typology of conservative, liberal, and socialist (non-Soviet type) analyses and proposed solutions.

The Facts

Amid a variety of attempts at a definition of hunger, the following seems adequate. Hunger is "inadequacy of dietary intake relative to the kind and quantity of food required for growth, activity and maintenance of good health" (Kates and Millman 1990, 3). Most important, hunger is generally involuntary and is usually chronic.

There is some variation in the levels of hunger. According to Millman, 20 percent of the world's population (1,053 million people) are receiving insufficient food energy for work (1991, 8). Nine percent of the world population (477 million people) receive "insufficient energy for the normal growth of children; minimal activity of adults" (1991, 8).

Referring to hunger as "food insecurity," Phillips and Taylor cite 1986 World Bank figures that "in 1980 between 340 and 730 million people suffered from undernourishment" (1990, 62). Regionally, "about two-thirds live in Asia and 20% in sub-Saharan Africa" (Phillips and Taylor 1990, 62). In the context of world population growth, Nevin Scrimshaw (1990) notes that "in the world today, more people suffer from malnutrition than ever before" (p. 353).

In addition to the lack of energy to work, malnutrition causes specific nutritional deficiencies. Before the twentieth century, these usually took the form of:

Beri-beri: a lack of B vitamins (often due to eating white, rather than brown rice)

Pellagra: a niacin deficiency (often due to eating a diet based on corn and sorghum)

Scurvy: vitamin C deficiency (often due to failure to consume enough fresh fruits)

Rickets: lack of exposure to adequate sunlight to convert ergosterol in skin to active vitamin D (Scrimshaw 1990)

Today's hunger-induced vitamin deficiencies often have some different effects. Generally malnourished or food-deprived people suffer from

Underweight: 24 million infants, who make up 16% of the world's infants. This results in stunted growth.

Underweight for age: 168 million children, who make up 31% of the world's children under five years.

Iodine deficiency: 210 million people, who make up 4% of the world's population. Most are Asian. Chen writes that "190 million suffered from goiter, the enlargement of their thyroid glands (endemic goiter), which may be accompanied by reduced mental function, lethargy, and increased fetal and infant mortality" (1991, 11). 3million suffer from cretinism, an irreversible neurological impairment (1991, 11). This also decreases resistance to infections.

Iron deficiency: 13% of the world population (700 million people). Scrimshaw (1990, 359) estimates this to be one-third of the population of developing countries. It means lowered learning ability, work performance, and resistance to infection

Vitamin A deficiency: Often coming in the aftermath of pervasive diarrhea in the developing world, vitamin A deficiency causes partial or total blindness, increases the "risk of respiratory infection and the incidence of complications from measles" (Chen 1990, 11).

The most extreme cases of hunger involve protein/calorie malnutrition, usually found in small children: "*kwashiokor* is the indigenous name used to describe the disease that occurs when a child is displaced from the breast by another baby. [It involves] moderate to severe growth failures and muscles that are poorly developed and lack tone . . . resulting in a large pot belly and swollen legs and face. The child has profound apathy and general misery; he or she whimpers but does not cry or scream" (Bryant 1985, 289–290). *Marasmus* is the lack of calories. It involves extreme growth failure and generally occurs in the child's first year. The child looks extremely emaciated.

Box 9.1 The Question of Breast Feeding

No one disputes the superior health benefits of mothers' milk over that of cows. As Sara Millman of Brown University's Alan Shaw Feinstein World Hunger Program writes: "Breast-feeding nourishes very young children, minimizes the exposure to environmental contaminants, provides some defense against infection and contributes to a relatively favorable pattern of birth spacing that in turn can have an important positive effect on the health of both your children and their siblings" (1986, 91).

In unsanitary and very low income conditions, breast feeding provides many immunities through antibodies from the mother. And it prevents the almost universal diarrhea in the developing world. This malady kills many babies through preventing the absorption of nutrients from food.

Unfortunately, the multinational oligopolies have sought a new market for their infant formula in the developing world. They employ radio and billboards. They make sure that samples of infant formula are dispensed in hospitals. And, in 1977 Nestlé Corporation was employing 4,000 to 5,000 "mothercraft advisors in nurse-like uniform" (Lappe and Collins 1977, 315). They were often given salaries and commissions.

This advertising has combined with the use of infant formula by many middle-class and working mothers to make it a status symbol of Western life. Just as eating at McDonald's is a status symbol for America's urban poor, so feeding one's baby on Western infant formula has become one mark of status in the developing world.

Unfortunately, the conditions of low-income life make infant formula a producer of death. If not used, a mother's natural milk dries up, necessitating the continued use of formula milk, and the expense of formula may take up to half a poor family's food budget. Because of the expense, uneducated families may dilute the formula with water to stretch it. This situation soon leads to serious malnutrition and death.

While infant formula instructions advise strict sterility, this is often not possible with polluted water supplies and limited cooking facilities. In addition, many of the developing world's poor are illiterate. Because they cannot read, they cannot follow the directions. And even if they could, they lack the education to understand the importance of a germ-free environment. In this context, infant formula becomes nutricidal.

In 1974 the British group War on Want began the international campaign against Nestlé with its pamphlet 'Nestlé Kills Babies" (Lappe and Collins 1977, 311). The worldwide boycott of Nestlé forced them to curtail this kind of infanticide. But recently they have started again, and the boycott has been reinstituted.

If, as Lappe and Collins claim, there is enough food for everyone, how does hunger persist? We must understand that, as Millman writes, "the history of hunger is embedded in the history of plenty" (1990, 3).

The overriding issue of contemporary hunger is that, as we noted in chapter 3, adequate nutrition can come from grains, but 40 percent of the grain grown in the world is fed to livestock to turn it into high-priced meat (*Politics of Food* 1987). In a world capitalist system, food goes to those who can afford to pay for it. Often they are the fast-food customers in the developed world. Only the socialist world seeks to provide for food as a right.

How did this situation come about? How does one value it? What, if anything, should be done about it? The following is a typology of conservative, liberal, and contemporary socialist analyses of and prescriptions for solving world hunger.

Conservative Analyses and Prescriptions

In *Food, Poverty, and Power*, Anne Buchanan reproduces a cartoon that reads: "Starvation Is God's Way of Punishing Those Who Have Little or No Faith in Capitalism." Conservative analysts of world hunger have generally seen a form of (Judaeo) Christian divine providence in the suffering of the poor ("The poor ye shall always have with you"—Matthew 26, 11). This is an ethnocentric position that generally values white Western life above all others and justifies the suffering from hunger on some combination of theological, value, or demographic grounds.

Garrett Hardin provided the classic demographic justification for doing nothing. His argument assumes an analogy between the earth and a lifeboat. Each has a limited carrying capacity. Overloading causes catastrophe for all. Therefore, those of us (in the developed world) who now control lifeboat Earth should not permit general improvement in the hunger situation. To do so would create too many people (overload the lifeboat) and decrease the quality of life we now live. Implied in this argument are ethical assumptions that the present world system is "right" and that we in the West, as its dominating force, have a responsibility to keep the lifeboat earth from overloading with (minority) races and classes.

Demographically, Hardin implores us to "guard against boarding parties" (1978, 75) that could take the form of the much higher reproductive rates of people in developing countries. His background is Thomas Malthus's 1798 work ('Essays on the Principle of Population'), which took the position that at the then current rate of growth, population would soon outstrip food production. According to Malthus, food production grows arithmetically while population grows geometrically.

In this context, Hardin argues *against* a World Food Bank creating a means of keeping people alive who might otherwise perish from "natural" causes. He detects the profit motive for private business in U.S. food aid efforts. In addition, he notes a desire for cheap labor as the motive in liberal demands for increased immigration quotas (1978, 79).

The second major conservative argument *for* the continuation of hunger and malnutrition is one that subverts the cultural relativism upon which liberal social scientists tend to pride themselves. That is to say, conservatives argue that peasant *attitudes* (values) are the source of hunger and malnutrition.

Hardin believes that the issue is one of "slovenly rulers" and incompetent governments. which do not plan for lean years. In *Famine in Peasant Societies* Ronald Seavoy continues this values-based argument. He characterizes the peasant mentality (value system) as "survivalist." Peasants work only to subsist (not to grow rich) and to allow for indolence. In fact, "peasants will not be motivated by commercial social values until political power is used to destroy the communal institutions that protect the subsistence compromise" (1986, 378).

Claire Cassidy (1982) proposes a similar value or culture-grounded argument in her analysis of the weaning customs of nonindustrialized society that may "potentiate" malnutrition. In the context of her general (liberal) concern for the hungry and malnourished, she shows that many peasant customs (such as early weaning) have a latent function of "potentiating" malnutrition while their manifest function is to socialize and display love for the child.

Cassidy also echoes Hardin's argument when she demonstrates that malnutrition serves to limit population growth and eliminate those biological organisms that are not strong enough to endure the malnutrient hardships of certain societies' food supplies. She writes, "the experience of malnutrition in early childhood may also be adaptive in the sense that it biases developmental plasticity toward the hunger-resistant" (1980, 333). She refers to this process (echoing Edward Banfield in *The Unheavenly City*) as "benign neglect."

While Hardin truly is a conservative, Cassidy justifies her excursion into conservative territory by her interest in providing more successful nutrition programs. According to her, if nutritionists understand the role of values in developing societies, they can structure really effective programs for alleviating hunger.

While Seavoy and Cassidy rightly explore the role of values in "traditional" societies, they both fail to come to terms with the etiology of values and the manner in which they change.

Value longevity has been the traditional conservative answer to most social customs. Banfield argues that the values of the lower class inhibit any kind of self-improvement. And those values are passed down through the "culture of poverty" in America. Such values as a "taste for excitement," "immediate gratification," and "timelessness" will go on forever among the poor. Therefore, one should do nothing. From a liberal materialist perspective, values are *responses* to material situations. That people in "traditional" societies hold many values in common is a collective response to the common conditions of precommercial, horticultural, or agrarian society.

Having large families is a good example. Seavoy argues that "large families are desired in peasant societies because child labor allows parents to enjoy indolence" (1986, 20). Leisure (values) motivate high birthrates. In contrast, materialists argue that high birthrates function in poor societies as a system of social security. By having a large number of (especially male) children, parents increase the chances of having an adequately supported old age.

Demographers speak of the "fertility transition." About forty years after a country "develops," the birthrate drops. From a materialist point of view, there is a new material reality to which to adapt. Developed societies generally have social security systems that replace the function of large families. In addition, developed societies often are predominantly urban. Children, who are positively functional for an agrarian society—they can do many farming tasks—are dysfunctional and an economic burden in the city.

From traditional sociology we can utilize the notion of "cultural lag" (William Ogburn) to explain the values aspect of this transition. Because values are an adaptation, they change slowly. Generally one's personal values are dominantly formed by the time of graduation from college. These values usually form the basis for the adult personality. But when material conditions change (a social security system is implemented; immigrants come to a different society), the second generation generally is the locus of a *change* in values. Having grown up with a social security system and an urban environment, it makes more adaptive "sense" to have (value) a small family.

As for Banfield's analysis of the "culture of poverty," how does one explain some tendencies to continue to have large families in America? First, many poor families in America are still rural. Extra hands are still needed on farms. And among America's urban poor, social security alone provides such a limited retirement income that there remains some belief that more children will help in one's old age. In fact, low-income families are more mutually supporting than those of the middle class (again, out of material necessity, more support is needed).

Summarily, conservatives like Garrett Hardin (1978) generally support hunger as useful in a potentially overpopulated world. They justify their position on grounds of protecting the existing elite and on grounds of the value systems of the malnourished and hungry. They "blame the victims" of poverty and malnutrition. The conservative solution to world hunger is to allow it to continue. While thinking of themselves as personally compassionate, conservatives support institutional indifference and benign neglect for the problems of the poor, hungry and malnourished. Their prescription for social change is to have none—even at the cost of 20 million lives a year.

Liberal Analyses and Prescriptions

Liberals traditionally tend toward technological explanations of history and political "reforms." Perhaps the best-known sociologists of technologically generated "evolution" are Gerhard and Jean Lenski (*Human Societies* 1982). They analyze human development from the perspective of food-growing technologies. They identify hunting and gathering, horticultural, agrarian, and industrial technologies of food production.

In the area of hunger, Esther Boserup provides the technological explanation. Blending demography with technologies of food production, she "claims that population growth itself stimulates agricultural innovation and leads to production increases that more than keep up with population growth" (Crossgrove et al. 1990, 224).

The technology of food production is the key. Instead of a contained "lifeboat" earth, the lifeboat may now be expanded as a function of a technological fix. As mentioned in chapter 8, the Green Revolution and biotechnology (and some birth control) are the current technological hopes to solve world hunger.

Like the conservatives, liberals assume the essential goodness of the world economic system. But, unlike the conservatives, they feel that with a bit of technological tinkering and policy reform, the world can alleviate food shortages.

The most important policy innovation is U.S. Public Law 480. Passing the U.S. Congress in 1954, PL 480 provides a foreign policy that disburses (surplus) food throughout the hungry world. It was the perfect liberal humanitarian concern of the United States for the world's hungry. Because the motives for "doing good" are often mixed, it also provided a solution to some economic problems of surplus U.S. food. In the Great Depression the United States dumped, burned, and plowed under surplus food rather than distribute it on a depressed market and therefore ruin the profits for existing businesspeople. With PL 480, this food could be disbursed abroad and at the same time profit U.S. producers. As Dan Morgan writes:

> PL 480 was advertised as an aid program for foreign countries, but above all it provided assistance to American farmers and the grain trade. Foreign governments received authorization from the U.S. government to purchase, with American loans, certain quantities of American farm commodities and the foreigners handled the actual transactions, contracting with private exporters to obtain the goods. But payment for these goods actually went straight from the U.S. Treasury to commercial banks in the United States and then to the private exporters.
>
> The foreign governments had the obligation to repay the loans, but the terms provided grace periods and long maturities. (1980, 147–148)

Implicit in virtually all the aid schemes of the United States and often of the United Nations is the assumption that world capitalism is the best system. Imperfections must only be reformed in order to have the whole system run better. Nevertheless, there are important criticisms of this approach. Most of these center on the political slant involved in the choice of aid recipients and the ineffectiveness of the aid in actually improving the hunger situation.

Probably every country counts as a side benefit of aid the creation of political loyalty and support. With PL 480, "the chief allocations go for political reasons to countries considered strategically important" (Kutzner 1990, 31). Lappe and Collins criticize U.S. aid programs for going disproportionally to politically loyal countries and having little to do with poverty. For instance, "Out of 70 odd governments receiving almost $34 billion in U.S. bilateral economic assistance in the first half of the 1980's, just 10 countries got over half the assistance . . . Israel and Egypt together got almost one-third (1986, 105).

Often this aid is dispensed to the most politically repressive dictatorships: "While El Salvador ranks third among per capita recipients of U.S. economic aid, its government defends the economic structures that have made Salvadorians among the five hungriest peoples in Latin America" (Lappe and Collins 1986, 107).

Similarly, food is used as a weapon. The withholding of aid is threatened or performed when countries institute policies that run counter to U.S. "development" objectives. Allende's Chile and Sandinista Nicaragua were but two countries that constructed social systems to feed their poor and develop means to feed them in the future. The United States cut off aid for both of them.

A second major criticism of U.S. (and world) food aid is that it is ineffectively done. Again, the assumption is the fundamental unchangeability of the capitalist market system. Only reforms that maintain profitability as a criterion are acceptable.

Aid often does not get to the people deserving of it. In many developing countries villages are controlled by local men (very few women) of power. In one example cited by Lappe and Collins:

> Thanks to a bribe to a technician, an irrigation pump earmarked for a cooperative of poor farmers in Bangladesh winds up belonging to the village's richest landowner: he graciously allows his neighbors water from the new well in exchange for a third of their harvests. And "... the landowner can now buy an imported tractor, eliminating desperately needed jobs for the village's landless families" (1986, 111).

Another form of ineffectiveness is the manner in which food aid can underprice local producers, driving them out of business. It can discourage local entrepreneurial and productive activity. It is essentially a "dole" on which people can become dependent. It relieves the local government of responsibility for providing structural changes that would actually make a given country self-sufficient in food (Lappe and Collins 1986, 11). And since much aid is in the form of loans, these countries often have to orient their agricultural production toward export crops.

Third, U.S. food aid may be either inappropriate or may function to change local food tastes away from indigenous sources of nutrients. Because the United States has often used wheat, there now exists some demand for it as an evidence of Westernization. In Africa, wheat grows poorly. As a result, a demand surfaces for imported wheat while many African countries cannot provide enough basic nutrients and have no foreign exchange with which to purchase wheat.

Until fairly recently, dried milk (of which the United States has a surplus) was often shipped to the African continent as food aid. Unrecognized was the pervasive lactose intolerance of most Africans (and many Asians). Because these people lack the enzyme lactase to break down the lactose, they suffered upset stomachs and diarrhea from the milk. Often the milk went unused or served as a whitewash.

India provides the best-known example of the multitude of liberal reformist approaches to decreasing hunger. Its focus has been on population limitation and technological improvements in food production. Twenty years ago, liberal American social scientists were advocating birth control as the solution to the Indian hunger problem. There were just too many people. At that time, a multitude of schemes were devised to provide incentives for Indians to use birth control and sterilization to limit births. Radios and other rewards were given out in villages as rewards. Not much was said of the violation of human rights by coerced sterilization programs.

Another area of liberal reform in India has been the attempt to increase food production. India provides one of the best examples of success in scientific agriculture. The genetic inbreeding (Green Revolution) of high-yielding varieties (HYV) after 1965 doubled India's grain production. However, these gains were only in specific regions: the Punjab, Haryana, and western Uttar Pradesh. They were in wheat more than rice. And they failed to increase the yield of jowar and bajra, the staple foods of the Indian poor (Hinrichs 1988, 7).

Furthermore, because of its skewed income distribution and lack of a national feeding program, "nearly half the population lacks the income necessary to buy a nutritious diet" (Lappe and Collins 1986, 50). In precisely the area of Green Revolution success, the ownership structure prevents the two-thirds of the population who are poor from eating successfully (Lappe and Collins 1986, 50).

In much of the developing world, when a surplus is generated, the owners sell it to the highest bidder. This often means that the owners of the grain export it, even in periods of massive famine. If the poor do not have the funds to purchase the grain, it is of no use to them in a capitalist system.

For this reason, many liberal attempts call for *entitlement* programs. They reason that everyone should be entitled to such necessities as air, water, and food. The problem is to accomplish this without intruding on the profit motive in a basically capitalist system. These programs often provide public jobs, which can then supply the cash to buy food. India has taken this approach. As long as the jobs are in the public sector, they do not compete with private profits.

Even in the mixed economy that existed under the Sandinista government in Nicaragua, ration coupons provided only for basic goods (rice, oil, salt), leaving the production and sale of more luxurious and varietous food commodities in private hands. As Dreze and Sen write: "Entitlement protection will almost always call for mixed systems, involving the use of different instruments to provide direct or indirect support to all vulnerable groups. The provision of employment—perhaps with cash wages—combined with unconditional relief for the 'unemployable' is likely to be one of the more effective options in many circumstances" (1989, 121).

Liberal analysts have had the main stage in hunger relief for a long time now. Yet hunger persists. Because of their commitment to a world capitalist system (however "reformed") and their usual optimism, they are perpetually proposing new reform "fixes." One recent reform was proposed by the previous chief economist at the U.S. Agency for International Development, John Mellor. Now a professor at Cornell University, Mellor calls for a massive public works program ($15 billion per year) and $5 billion per year for feeding programs. He asks that the World Bank and World Food Program join to spearhead this effort (Mellor 1990, 499–500). He wants serious money spent on rural infrastructures that would allow "development" and "modernization" to occur. Should that happen, countries could begin to become self-sufficient in food.

On the other hand, Mellor recognizes the utopian nature of his scheme. He writes: "the politics of foreign aid and national allocations have determined much of the present allocational pattern and it seems unreasonable to expect major changes to occur in the future" (1990, 507).

In sum, liberal analyses and plans abound. Most are well intentioned. Yet their track record, as demonstrated by the country in which they have been most thoroughly applied, India, is short of middling. Without changing the economic system that has and continues to create hunger conditions, there is little real hope that hunger will be ended. Only in the contemporary socialist world has hunger been effectively eliminated. It is to that world that we now turn.

Political-Economic Modern Socialist Analyses and Prescriptions

By way of introduction, I want to make it clear that a socialist analysis of the world hunger problem is not an analysis done from the perspective of the former Soviet Union or eastern block. It has very little to do with the "state capitalist" or "coordinator" societies that existed in eastern Europe before the current movement toward capitalism.

Instead, a modern socialist analysis partakes of the intellectual tradition of Western Marxism. This has been variously called neo-Marxism, critical sociology, dialectical, materialist, political economy, or "radical" analysis. All relate to an intellectual position where one looks at the political structure in relation to the economy before one looks at societal norms, ideas, values, and ideologies. This perspective assumes the more important aspects of the social system are to be found in the qualitative analysis of the economy and the manner in which politics and ideology generally flow from it.

In the area of world hunger, this position has been termed neo-Marxist dependency theory, political economy, world systems theory, and radical democratic theory. In all cases, it locates the reasons for hunger in the functions of the economic system.

This economic system is composed of two parts: ownership patterns and technology. In the world capitalist system, the basic ownership pattern is that of private capitalist elites owning most of the productive apparatus in the world. To them, in one form or another, workers (including farmworkers) sell their labor power.

This book has already alluded to some of the technology involved in world hunger. Primary are those that relate to food production. Chapter 8 provided a more thorough and specific analysis of the variety of technologies available to produce food.

Suffice it to say that in a predominantly capitalist world system, the oligopolies own and control most of the significant productive apparatus that relates to food. Technologies include land, farming machinery, farm chemicals, seeds, and knowledge. Oligopolies also bias toward their interests the research on the future profitable technologies of food production.

A situation in which one group has essential control of a process (but is not totally in control) is called *hegemony*. Small farmers in America can produce what and how they want, but the oligopolists control most large-scale production. And because of economies of scale, they can usually run small farmers out of business. Only certain specific crops (including many vegetables) that require highly labor-intensive production escape this situation at the moment. Future technology may even obviate this labor need.

From a historical perspective, the problem of food poverty and hunger is rooted in colonialism. Most indigenous peoples knew how to produce enough food for themselves. The groups that did not did not survive to tell about their nutritional or production mistakes.

The eighteenth century marked the encircling of the globe by new capitalists. England, France, and Spain staked out claims to a large part of the world's land. Though this process was often performed under the ideology (belief system) of bringing Christianity and Western civilization to the "heathen," in fact, it brought mountainous profits to the entrepreneurs who invested in it. Some critics of colonialism are fond of saying that the missionaries brought the Bible to the new lands. When they were through, the "natives" had the Bible and the missionaries had the land.

This modern socialist perspective conceives of the government (or state) as bowing to the demands of its leading (socioeconomic) class. These are the capitalists. They provide the funds to finance elections of people who represent their interests and who would goad their governments to provide legitimacy and protection for their foreign ventures. These "democracies" were established to appear democratic but in fact act in the interests of the capitalist classes.

In what is now the developing world, the plantation pattern often dominated. For instance, in the English Caribbean, sugar became the principal product returned to England. It both enriched the capitalist class and impoverished the diet of the English industrial working class (Mintz 1985 and chapter 3). Government-sanctioned slavery created much of the early capital by means of the triangular trade in sugar.

The nineteenth century brought many struggles for political independence throughout the now developing world. Because these subsistence economies had already been reorganized as plantations, however, it did not bring a return to subsistence economic production. Instead, the pattern of ownership switched from colonialism to neocolonialism. Instead of having British administrators (as in India), local governments ruled in name while private foreign companies dominated the economies, especially the agricultural sector.

Instead of producing food to feed their populations, foreign countries often employed local administrators to produce export crops. Tea (in India) and sugar (in the Caribbean) continued to be sent back to the imperial countries for processing.

Thus developed a division of labor that has characterized relations between the developed and developing world to this day. Raw materials (which have lower profits) come from the developing world. The developed world imports raw materials and processes them. It is in the latter process that the greater share of profit is to be found.

Andre Gunder Frank (1969) termed this process "the development of underdevelopment." By it, the developed world essentially inhibits indigenous development and keeps the developing world producing raw materials for export. And, in those instances where food crops are processed locally, generally the processing companies are owned by foreign stockholders and the profits repatriated to the developed world.

The new development in the twentieth century has been the growth of multinational world oligopolies in food. "Companies such as Cargill, Continental Grain (both American) and Bunge (Argentina), Dreyfus (France) and Andre-Garnac (Switzerland) have an almost total monopoly of the U.S. grain trade" (Bennet and George 1987, 177).

Of greatest importance is recognition of a pervasive pattern in the developed world's treatment of colonial and subsistence food production patterns. The most extensive treatment of this pattern is to be found in the literary production of the Institute for Food and Development Policy headed by Frances Moore Lappe and Joseph Collins. Their 1977 *Food First* traced the development of this pattern in country after country throughout the developing world. We shall refer to this analysis as the *Lappe-Collins paradigm.*

Therein, contemporary multinational oligopolistic capitalist corporations enter a developing country in search of what all capitalists seek: high profits. Because most of the developing world is not highly industrial, the choice mode of profit generation has been the agricultural sector.

Whereas before the oligopolists' entry, peasants may have been successfully producing the subsistence crops that they had always eaten, now the newcomers gain control over the land on which this production has taken place. In the best of cases they may buy it. Sometimes they simply employ local thugs to murder or move the owners (*Politics of Food* 1987). Regardless, the company gains control of the land.

Because the world market is their target, they decide to grow whatever agricultural commodity will sell the most profitably on the world market. It may be flowers for Europe, coffee for America, or bananas, strawberries, sugar, vegetable oils, or oranges.

A most important need in the developed world is grain to feed cows. The beef of the developing world is a major part of the American fast-food industry (Heller 1985). Because land in the developing world comes relatively cheap, there is little ecological concern. Although it is true of capitalist agriculture generally that it has tended to ruin the land (see chapter 8), land in the developing world is even more exploited.

In Costa Rica, a leading exporter of beef to the United States, land is overgrazed and therefore ruined. As a result, many administrators and local owners are cutting rain forests to convert them to grazing land. Rain forests contain a large number of potential medical substances that have yet to be explored. Franke and Chasin report that one-quarter of Central American rain forests have been destroyed and only half a million of several million plant species have been saved: "The relationship between ecological destruction and food production is . . . direct and close. Whenever an environment is degraded, deprived of its basic resources—or even *one* of the key resources—that environment becomes part of the world food crisis, and the people who live there become its victims" (Franke and Chasin 1980, 4).

The lot of the original farmers is even more discouraging. Whereas a country may have previously had a stable rural laboring population, now they are displaced. If they choose to stay on the land, some of them are usually hired back to work for the corporation that now owns it. Because they no longer produce their own food, they must purchase the food they previously produced. This practice often gets them further into debt. It is reminiscent of the system of debt peonage in the United States after the formal demise of slavery.

Because food production methods in capitalist developing countries are generally high technology (capital intensive), fewer people are needed to work the land. Those for whom there is no work often flee to the urban areas where they become part of the "informal" urban economy of beggars, street peddlers, and thieves. Capitalists have always relished a surplus of potential workers. In a market for labor power, more people competing to sell their labor power means that the labor price can be significantly lower.

The Lappe-Collins paradigm takes a variety of forms throughout the developing world. Different crops require different soil conditions. Sometimes the political structure is controlled to provide protection for capitalism (Chile after Allende and Nicaragua after the Sandanistas). Sometimes there is a need for more labor-intensive agriculture. But, as Lappe and Collins documented in their voluminous *Food First* (1977), it applies as a paradigm in almost every country in the developing world.

In this context, food and financial aid are viewed somewhat more cynically. Unlike the liberal position, which understands aid as potentially helpful, modern socialist analyses are suspicious of Western aid. They note the political and economic debts that come with aid. And they note the manner in which such institutions as the World Bank generally sponsor agricultural endeavors that fit the Lappe-Collins paradigm of cash crops for export. This usually breeds dependency, poverty, and hunger in the developing country.

Modern socialist prescriptions vary. But generally they include food, security, and empowerment. They want people to be able, with the aid of contemporary technology, to grow food for sustenance needs. They want to stop ecological destruction, especially of rain forests and areas adjacent to deserts. They want to develop local industry to process their own agricultural produce. If cash crops are produced and sold on the world market, socialists want the money to be used for the development of their country, rather than shipped to the developed world or used by developing elites for luxuries.

Generally, modern socialists propose appropriate technologies. Instead of importing the technology developed for capital-intensive farming, modern socialist countries often prefer labor-intensive farming. Such social technologies employ people and allow ecological farming practices such as intecropping and non-chemical pest control.

These principles have guided food production such in places as China. As opposed to India, China has for a number of years mostly eliminated hunger. Though liberals may have qualms about China's authoritarian family planning, political regime, and move toward capitalism, the fact remains that in China a lesser proportion of the population is undernourished than in the United States (Hinrichs 1988, 14). As Harrison notes, "China has, by all accounts, solved its food problem" (1980, 150). In comparison to India's death rate of 12 percent per thousand, China's death rate of 7 percent per thousand leaves India with an "excess normal mortality . . . of 3.9 million per year" (Dreze and Sen 1989, 214).

Summary

In retrospect, conservatives look on world hunger as natural and, in a sense, good. Like evolution, it weeds out the weak and rewards the strong. If people in the developed world begin with a significant head start, they are not to blame for that.

The liberal position is reformist and well intentioned. It wishes that the developed world would become altruistic and contribute enough money and social technology to bring an end to hunger. Its heroes are the U.S. Agency for International Development and the World Bank. Only those organizations are big enough to fund change.

But the liberal position overlooks the stake the capitalist multinational oligopolies have in underdevelopment and dependency. Profit maximizers need hungry and needy people for their low-paid workforces. They need cheap land that can be inexpensively farmed and where one need not take ecological considerations too seriously. And they need the multitude of dictatorial regimes that the United States supports throughout the world to keep their countries safe for exploitation by multi-national corporations of the developed world.

The modern socialist position is the only one that provides an adequate analysis of world hunger. But a solution, according to this analysis, would involve a socialist government. China, Cuba, North Korea, Sri Lanka, and the Indian socialist state of Kerala provide examples of wide-scale success in solving hunger. It is to them that we now turn.

References

Anand, Sudhir, and S.H. Ravikanbur. 1991. "Public Policy and Basic Needs Provision: Intervention and Achievement in Sri Lanka." In *The Political Economy of Hunger*, vol. 3, *Endemic Hunger*, ed. Jean Dreze and Amartya Sen, 59–92. Oxford: Clarendon Press.

Barkin, David, Rosemary L. Ball, and Billie R. DeWalt. 1986. *Food Crops vs. Feed Crops*. Boulder, Colorado: Lynne Reiner.

Basu, Kaustrick. 1991. "The Elimination of Endemic Poverty in South Asia." In *The Political Economy of Hunger*, vol. 3, *Endemic Hunger*, ed. Jean Dreze and Amartya Sen, 347–371. Oxzford: Clarendon Press.

Bennet, Jon, and George, Susan. 1987. *The Hunger Machine*. Toronto: CBC.

Bread for the World. 1992. Washington, D.C.:Institute on Hunger and Development.

Bryant, Carol, Anita Courtney, Barbara A. Maikosbery, and Kathleen DeWalt. 1985. *The Cultural Feast*. New York: West.

Buchanan, Anne. 1982. *Food, Poverty and Power*. Nottingham: Spokesman.

Cassidy, Claire. 1982. "Protein-Energy Malnuitrition: A Culture Bound Syndrome." *Culture, Medicine and Psychiatry*, *6*:325–348.

Chen, Robert. 1990. "The State of Hunger in 1990." In *The Hunger Report: 1990*. Providence, R.I.: Alan Shawn Feinstein World Hunger Program, June, 1–26.

Cheng, Wei-yuan. 1986. "Testing the Food First Hypothesis: A Cross-National Study of Dependency Sectoral Growth and Food Intake in Less Developed Countries." Paper presented at the meeting of the Association for the Stufy of Food and Society. Grand Rapids, Mich., April 22–24.

Crossgrove, William, Peter Heywood Egilman, and Jeanette Kasperson. 1990. "Colonialism, International Trade and the Nation State." In *Hunger in History*, ed. Lucille F. Newman, William Cosgrove, and Robert Kates, 220–2312. Cambridge, Mass.: Basil Blackwell.

Dreze, Jean, and Amartya Sen. 1989. *Hunger and Public Action*. Oxford: Clarendon.

Fieldhouse, Paul. 1986. *Food and Nutrition: Customs and Culture*. New York: Croom Helm.

Foster, Phillip. 1992. *The World Food Problem*. Boulder, Colo.: Lynne Reiner.

Fowler, Carey, and Pat Mooney. 1990. *Shattering: Food, Politics and Loss of Genetic Diversity*. Tucson: University of Arizona Press.

Frank, Andre Gunder. 1969. *Latin America: Underdevelopment or Revolution*. New York: Modern Reader.

Franke, Richard W. 1987. "The Effects of Colonialism and Neocolonialism on the Gastronomic Patterns of the Third World." In *Food and Evolution*, ed. Marvin Harris and Eric B. Ross, 455–480. Philadelphia: Temple University Press.

Franke, Richard W., and Barbara H. Chasin. 1980. *Seeds of Famine: Ecological Destruction and the Development Dilemma in the West African Sahel*. New York: Universe Books.

Grant, James P. 1987. *Famine Today = Hope for Tomorrow*. Providence, R.I.: Alan Shawn Feinstein World Hunger Program.

Hardin, Garrett. 1978. "Lifeboat Ethics: The Case Against Helping the Poor." In *The Feeding Web*, ed. Joan Dye Gussow, 74–85. Berkeley, Calif.: Bull Publishing.

Harrison, Gail Grigsby. 1980. "Strategies for Solving World Food Problems." In *Nutrition, Food and Man*, ed. Paul B. Pearson and Richard Greenwell, 141–152. Tucson: University of Arizona Press.

Heller, Peter. 1985. *Hamburger I: Macprofit*. New York: Icarus Films. Videorecording.

Hinrichs, Cynthia Claire. 1988. "Attaining Food Self Sufficiency: The Contrasting Cases of India and China." Paper presented at the Association for the Study of Food and Society, Chevy Chase, Md., May 27–29.

Hoogterp, Bill, Jr., and Jason Lejonvarn. 1990. *Hunger and Homelessness Action*. Minneapolis: Campus Outreach Opportunity League.

Kates, Robert W. 1987. *Ending Hunger for the Second Billion*. Providence, R.I.: Alan Shawn Feinstein World Hunger Program, August.

Kates, Robert W., and Sara Millman. 1990. "Ending Hunger: The Lessons of History." In *Hunger in History*, ed. Lucille F. Newman, William Cosgrove, and Robert Kates, 389–415. Cambrdige, Mass.: Blackwell.

Kutzner, Patricia. 1990. "World Hunger: What Have We Learned?" *Hunger Notes 16*:1–2.

Lappe, Frances Moore, and Joseph Collins. 1977. *Food First: Beyond the Myth of Scarcity*. Boston: Houghton Mifflin.

———. 1982. *World Hunger: 10 Myths*. San Francisco: Institute for Food and Development Policy.

———. 1986. *World Hunger: Twelve Myths*. New York: Grove Press.

Lenski, Gerhard, and Jean Lenski. 1982. *Human Societies*. New York: McGraw-Hill.

Mellor, John W. 1990. "Ending Hunger: An Implementable Program for Self Reliant Growth." In *The World Food Crisis*, ed. Hans J.I. Baaker, 485–518. Toronto: Canadiam Scholars Press.

Millman, Sara R. 1991. *The Hunger Report: Update 1991*. Providence, R.I.: Alan Shawn Feinstein World Hunger Program, April.

Millman, Sara R., and Robert Wikates. 1990. "Toward Understanding Hunger." In *Hunger in History*, ed. Lucille F. Newman, William Cosgrove, and Robert Kates. Cambridge, Mass.: Basil Blackwell.

Mintz, Sydney W. 1985. *Sweetness and Power*. New York: Viking Penguin.

Morgan, Dan. 1980. *Merchants of Grain*. New York: Penguin.

Nair, K.N. 1987. "Animal Protein Consumption and the Sacred Cow Complex in India." In *Food and Evolution*, ed. Marvin Harris and Eric B. Ross, 445–454. Philadelphia: Temple University Press.

Nef, Jorge, and Jokelee Vanderkop. 1990. "Food Systems and Food Security in Latin America." In *The World Food Crisis*, ed. Hans J.I. Baaker, 97–138. Toronto: Canadian Scholars Press.

Peassa, Anal. 1983. *Trends in Food and Nutritional Availability in China*. Working Paper 607. Washington, D.C.: World Bank.

Phillips, Truman P., and Daphine S. Taylor. 1990. "Food Insecurity: Dynamics and Alleviation." In *The World Food Crisis*, ed. Hans J.I. Baaker, 61–96. Toronto: Canadian Scholars Press.

Politics of Food, The. 1987. Yorkshire, England: Yorkshire Television. Videorecording.

Riskin, Carl. 1990. "Food, Poverty and Development Strategy in the People's Republic of China." In *Hunger in History*, ed. Lucille F. Newman, William Cosgrove, and Robert Kates, 331–353. Cambridge, Mass.: Basil Blackwell.

———. 1991. "Feeding China." In *The Political Economy of Hunger*, vol. 3, *Endemic Hunger*, ed. Jean Dreze and Amartya Sen, 15–58. Oxford: Clarendon Press.

Scrimshaw, Nevin. 1990. "World Nutritional Problems." In *Hunger in History*, ed. Lucille F. Newman, William Cosgrove, and Robert Kates, 353–373. Cambridge, Mass.: Basil Blackwell.

Seavoy, Ronald. 1986. *Famine in Peasant Societies*. New York: Greenwold Press.

Stein, Stanley J., and Barbara H. Stein. 1970. *The Colonial Heritage of Latin America*. New York: Oxford University Press.

USDA. 1985. *Agricultural Outlook*. Washington, D.C.: Government Printing Office, December.

Yanqui Dollar. 1971. New York: NACLA.

Chapter 10

Food Success Stories

Much of this text has detailed a long line of mistakes and failures with food. It often falls to the role of sociologists to debunk and deconstruct society. Food maladies are no exception. Yet some examples of successes in food do exist. They fall under two categories: poor countries that have succeeded in feeding all their people, and changes in diet in the developed world toward better nutrition. In addition (see Appendix) we know of at least one instance in which food has been used self-consciously as a form of psychosocial therapy.

The examples of successful feeding come from the previously underdeveloped (or developing) world. They are some of those countries that have chosen a socialist path to development. China is the most important example of a large country. Cuba and Sri Lanka supply a success model among small countries. Even the Indian state of Kerala provides some accomplishments as a socialist organization within a capitalist country.

In the developed world, some successes stem from the influence of scientific nutrition combined with the effects of the "counterculture" in creating the health and organic food movement. In the context of massive advertisement of junk food, this successful influence on the eating habits of a large number of Americans is surprising.

Everyone Eats: China, Cuba, Sri Lanka, and Kerala

Certain characteristics unite the modern socialist approach to food and define it in opposition to most capitalist practices.

The *right* to food is basic. Unlike the capitalist world where food is rationed by price, the modern socialist world rations food on the basis of approximate equality and need.

This practice has its problems. Rationing can be bureaucratically cumbersome. A black market often coexists with the ration system. Lines, sometimes very long, become a part of everyday food life. And ration stores may not always have sufficient food to meet "consumer" demand.

On the other hand, modern socialist countries do not have the phenomenon of homelessness. People, in addition to having basic lodging, do not sit outside well-stocked food stores because they cannot afford to eat: "While consumers in these countries complain about the lack of variety of foods . . . everyone has access to an adequate nutritional diet" (Warnoch 1987, 132–133). Relative to the norms of their societies, there is no poverty.

In these developing countries, planning replaces the market. In production and distribution, decisions are consciously made with an intended result. But not all intended results are beneficial. China's Great Leap Forward was a massive failure in agricultural production. Nevertheless, a belief exists in the possibility of rational planning as an alternative path to development in food.

Additionally, most modern socialist countries have carried out some kind of land redistribution or reform. Because the society that preceded theirs was usually feudal or neocolonial, land was often concentrated in the hands of large lot owners. In order to empower the general populace, land had to be reallocated. In planned economies, the amount and type of land reallocated in this manner has varied—often according to political swings that sometimes have been extreme. Nevertheless, the ownership of land has been an important issue in the creation of these socialist economies.

Finally, the technologies of both labor organization and farming are usually "appropriate." This is to say, they are unfettered by influential multinational oligopolistic corporations seeking to maximize short-term profit. Modern socialist governments can choose technologies with consideration for traditional techniques, reasonable new techniques, and some degree of democratic participation in the choices and monitoring of what is chosen.

China[1]

In comparison with India, China has generally had a much more successful route to development. On just about all the important indices, China is way ahead: life expectancy (almost 70 years to India's upper 50s), under age mortality (47 per 1000—about a third of India's), and percentage of infants with low birthweight (one-fifth of India's rate). China is far ahead of India in general health and nutrition (Dreze and Sen 1989, 204).

Most of this successful development stems from socialist principles adhered to after the 1949 Chinese Revolution. The first Chinese land reform redistributed land from the rich peasants to the poor and middle peasants. This was about 50 percent of existing land. But by 1956, almost 90 percent of households belonged to large cooperatives (Warnoch 1987, 276). The years 1958 to 1960 marked the disastrous communes that attempted to allocate income on the basis of need rather than work. New Chinese policy rewards intensive farming on small plots of land and individual initiative. Although capitalist, it is hardly the capitalism of oligopolies and multinational corporations.

Most important, through all the changes of plans, has been the Chinese commitment to the general nutritional welfare of all. As Warnoch writes:

> Household consumption data for the 1960s and 1970s indicated that the ratio of the top 20% of income earners and the bottom 20% was the lowest

> in Asia, including Japan, South Korea, and Taiwan. Paine (1976, 291) concludes that per capita consumption to 1975 increased at a rate of around 2.4% per annum. Furthermore, this was on top of a *consumption floor* that included universal health and education, a guaranteed level of consumption of food, cheap housing, rationing of scarce goods when necessary, and a system of relative prices that discriminated against luxury goods (1987, 279).

In China, people get their food in a number of different ways. Basic rationed goods such as cereal and cooking oil are purchased with residence cards at one's individual consumer unit (Warnoch 1987, 279). In addition, state-run stores sell groceries more or less at a market price. Further, individual links can be established with countryside growers. And there are street vendors throughout Chinese cities.

The pattern is basic to modern socialism. Fundamental nutrition is guaranteed, but the price goes up for luxuries. Everyone eats, but those who are industrious (or connected to the party) often eat fairly well.

In chapter 9 we described China's use of appropriate technologies. With problems of overpopulation and no oligopolies, the technology of choice has been labor-intensive, traditional, and somewhat ecologically sensitive. One might in fact see China's model, combined with some appropriate biotechnology, as applying to a lot of the developing world. Anderson writes: "The future of humanity probably depends on combining Chinese-type intensive agriculture with the techniques of the 'high tech' era" (1988, ix).

Finally, the quality of the Chinese diet holds some lessons for the developed world. The comprehensive "'China Project' has collected 367 items of information each on over 6,000 people, generating more than 100,000 correlations between lifestyle and about 45 different diseases" (Liebman 1990b,

1,5–7). One director, T. Colin Campbell, a nutritional biochemist at Cornell University, reports that preliminary conclusions show that a "diet made up of at least 80%–90% plant materials may be optimal" (Liebman 1990b, 5–7). In contrast, the American "diet averages only about 30% plant materials" (Liebman 1990, 5-7).

This way of eating provides for all the essential nutrients while remaining low in cholesterol and salt. And Leung notes that "one of the health advantages that China has over capitalist underdeveloped countries was the absence of advertising of 'junk' foods by multinational corporations" (in Warnoch 1987, 143).

In sum, the Chinese system manifests few of the problems of the colonized developing world. By going it mostly alone, China tapped the wisdom of its traditional farming and dietary practices to feed all its people. While critics may legitimately complain about the travesty of Tiananmen Square, let us not ignore the massive successes of China in food and agriculture.

Cuba

Both Cuba and Sandinista-led Nicaragua have been portrayed as serious threats to the United States. While it is true that Cuba once posed a military threat (the Cuban missile crisis), the real threat of both countries is as alternative models of development. From the University of Michigan, John Vandemeer wrote in 1986 that, like Cuba, "free Nicaragua threatens the U.S. system in the same way that the ideas of the abolitionists threatened the southern plantation system" (p. 17).

Even though Nicaragua was a distinctly mixed economy (capitalist and socialist), it was important that the United States devote large expenditures to supporting the *Contra* fighters and the Chamaro party at election time. Now that Nicaragua no longer claims an affinity to socialism, the United States seems to have put its welfare on the back burner.

Nevertheless, Cuba could pose a major threat. For in the three decades since its revolution, Cuba has become a humane, seriously democratic (even Fidel Castro is now elected) country that has seen much organizational and technical innovation. Its neighborhoods revolve around People's Power (*Poder Populair*) in the legislative arena and the Committees for the Defense of the Revolution (CDRs) in the self-help and maintenance areas. Where else in the world do people sweep their streets on Sunday, help recycle all neighborhood trash, aid in birth control and school problems, mediate neighborhood disputes, and elect neighborhood people who are really known (and immediately recallable) to district and state governing bodies?

Part of the success of the Cuban model has been in food and agricultural policies.[2] Like the rest of the modern socialist world, Cuba's commitment is that everyone eats. And their agriculture provides full employment for a multitude of workers.

Cuba uses rationing. Though they would prefer to avoid it, there seems no alternative and it is both fairer and speedier. Included among the foods viewed as a basic right (and therefore rationed) are "rice, beans, oil, sugar, salt, coffee, meat, chicken, and some fruits" (Benjamin et. al. 1984, 28). Eggs, spaghetti, fish and butter were plentiful and cheap enough to come off the ration in 1984.

Each family has a ration book and must always shop at the same government-owned store. Special allowances are given to senior citizens and mothers with small children. Special medical needs are also taken into account. People go to the stores with their containers in shopping bags. Each should have a bottle or can for cooking oil, baby food jars, and something in which to carry lard. The bags are not disposable paper but reusable straw.

Critics have complained about the lines that develop at government stores. While this is true, there are some latent benefits and some adaptations for two-parent working families.

From a neighborhood point of view, the lines become a matrix for political discussions about community problems. They are a sustained chance for neighbors to know each other and exchange views. This neighborhood interaction is precisely the desire of many urban sociologists when they speak about the networks necessary for safe neighborhoods.

To accommodate the increase in two-parent families where both parents work, *Plan Jaba* began in 1971. A family member will bring the straw bags in the morning with the ration book and return at night to pick up the bags, which the store manager has filled.

An additional source of nutrition in Cuba is the restaurant. Going out to dinner can be a source of entertainment and (with long lines) a source of frustration. Nevertheless, one accepted practice in Havana is to, by oneself, order three meals. Then, when the waiter has turned away, to slip these into a plastic bag to take home to eat during the week.

Much has changed since the collapse of the USSR and its substantial subsidy. Now most food is rationed. Cubans generally get only two full meals a day, and restaurants are limited to people with "hard" currency. Without the Russian subsidy and still subject to the U.S. embargo, Cuba is barely hanging on.

In agriculture, there were agrarian reform laws in 1959 and again in 1962. These coincided with the breakup of some of the large estates on which sugar was grown. And they supported a policy of agricultural diversification, which Cuba attempted early in its revolutionary program. Despite initial attempts at diversification, Cuba settled on (or was pushed into) Soviet subsidization and a place within the Eastern Block division of labor, which required Cuba to maintain its main crop as sugar.[3]

The position in the international division of labor that required Cuba to grow sugar necessitated the maintenance of large agricultural units. Sugar grows best in the tropics. One result of the concentration on sugar has been the invention (and now partial export) of a mechanical sugar harvester. By 1983, 60 percent of the sugar harvest was mechanized. As Benjamin et. al. write: "Cuba is now the largest producer of cane combines in the world" (1984, 134).

In agricultural labor organization, Cuba attempts to involve all its people and tailor that part which is mechanized to the hardest jobs. In an admittedly ideological attempt to overcome divisions between city and countryside, many Cuban students and office workers do some agricultural work. Students may labor as part of an ordinary school day. Many office workers labor during the peak harvest season.

With current food shortages, the Cuban goverment has encouraged widescale gardens and even urban chicken and pig raising. Because Cuba cannot afford petrochemicals, it has quickly moved to the forefront of organic farming. Unfortunately, Cuban culture devalues vegetables as food for rabbits!

Probably Cuba's weakest point, from a nutritional point of view, was its attempted imitation of Western high-fat, high-sugar diets. This is, from one perspective, understandable. In most of the underdeveloped world, fat people eat meat and set the standard of beauty (see chapter 5) because they are prosperous. As Cuba fed all its people, there were increasing demands for more meat. In fact, one of the general compliments in Cuba today is "*Que gorda estas*" (how fat you are!). To say this in the developed world context would certainly not be viewed as a compliment. Cubans traditionally have devalued raw vegetables, but place high value on sugar and coffee. Much food is fried. But the current agricultural and food crisis may lead to cultural changes.

In short, a kind of *nouveau riche* approach to eating, combined with traditional tastes, has pushed Cuba toward the vascular diseases of the developed world. Ironically it is their traditional rice, beans, and corn that is a better solution to nutritional balance. The withdrawal of the Soviet subsidy has forced some return to their original diet.

In spite of its flaws, Cuba has provided a successful example of one manner to overcome hunger and poverty through a firm adherence to the socialist path. Much of the mainstream reporting about Cuba still bears the marks of the Cold War. This author's trip there in 1977 found an open society with none of the heavyhandedness that characterized "socialist" eastern Europe. And precisely because of Cuba's successes, I suspect, the U.S. government does not permit Americans to travel freely there.

Returning to Cuba in 1993, I found the withdrawal of the Soviet subsidy serious. In this "special period in time of peace," Cubans are suffering. But because the suffering is equitably shared, there is no strong support for a change in government or social systems.

Cuba is now experimenting with the combination of traditional farming knowledge (from their older farmers) and biotechnology. Without petroleum sources of substance, they have generally returned to animal sources of locomotion for plowing and transportation of farm products.

Cuba's threat is that of a socialist technology firmly grounded in the ecological and food needs of the whole country. They use very few petrochemicals. This makes them an enormous experiment in sustainable, mostly organic, food production. Their survival depends on it.

If they succeed, they will again threaten the petrochemically based world agricultural system—especially in U.S. dominated Latin America—with a possible alternative path to development.

Kerala

Of special importance is the Indian state of Kerala. Though showing lower income averages than other states in India, "the state's . . . people are better educated and have better access to health care than almost any population" in the developing world (Franke and Chasin 1990, 29–30).

Kerala accomplished this through 9,227 feeding centers and fair-price shops that provide basic rations and 4,977 official village libraries (Franke and Chasen 1990, 29–30).

Kerala has more extensive medical facilities than other Indian states. In general education, "Kerala has been well ahead of the rest of India since at least the beginning of the 20th century" (Franke and Chasin 1990, 51). Much of its success stems from an extensive land-reform program carried out in 1969.

In Kerala and in places such as Sri Lanka, the government has intervened to provide necessities for the general population. Unlike the false claims of "trickle down" from "development," Kerala and Sri Lanka have provided real support for the advancement of the general population.

Other Socialist Examples

Among the other modern socialist approaches to food, one should at least notice Nicaragua (Sandinista), North Korea, and Sri Lanka. All are modern socialist approaches to food that have generally succeeded in feeding their entire populations.

Common to them are the elements of planning, government food stores (and the lines that are often involved in some degree of rationing), equalization of land holdings and some higher concentration on appropriate technologies (including labor-intensive) for farming.

All of these social experiments take advantage of the nutrition in their native diets to provide the basic nutrients needed. And all have a commitment to feeding their entire populations as a right rather than in response to ability to pay.[4] Though not all follow the new scientific nutritional lines, they have essentially overcome the hunger that existed in their countries. That people do not starve to death is one benchmark by which to judge the success of any country.

The Food Counterculture and Its Penumbra

At the present time, eating in the United States stands at the intersection of two dominant currents: the insights of nutritional science and the continuation of what was the "counterculture" of the 1960s and 1970s. Their interaction accounts for much of the change in American eating preferences in the 1980s and 1990s.

In the past decade we have witnessed an explosion in reporting the findings of nutritional science. Once obscure by mass reporting standards, nutritional science has become a hot topic in the popular and academic press. It is now common to have articles on health and nutrition in a wide variety of popular publications from news weeklies to check-out-line sensationals to the daily press. However, this reporting usually is minor compared to the advertisements in these same publications for the junk food that provides the high profits for the mainstream of the American food industry.

In spite of this inequality of print space, one still must notice that there have been some rather significant changes in the American diet. For instance, Brewster and Jacobson report: "from 1976–1981, coffee and tea use was down 17%. Palm oil was down and sunflower oil was up. Rice usage was up 55%. Meat was down 20% and poultry use was up 20%" (1983, 77-83).

In addition, Brody (1987) reports that the USDA National Food Consumption Survey reported a change in women 19–50 years old from 1977 to 1986. Their consumption evidenced "a decline in meat and pork consumption, a decline in eggs, an increase in persons using low fat skim milk, a decrease in whole milk usage, an increase in fish usage and increased fruit consumption" (1986, 13). Bill O'Neil, spokesman for the trade publication *The Packer*, reports in the *Grand Rapids Press* that consumption of fresh fruits and vegetables went from 72 pounds in 1970 to 102 pounds in 1989.

Among the big gainers in vegetables for the top 40 percent of consumers is broccoli, "up 214% between 1973-83, to 2.2

pounds per capita" (Belasco 1989, 219). According to Belasco, "for the non-healthy 60%, the most popular leafy green vegetable remained iceberg lettuce which, lacking nutrients or intrinsic taste . . . served mainly as a vehicle for dressing rich in salt, fat and calories" (1989, 219).

Most interesting is the division of the market by income and (most probably) education. For a long time now, the secularization of the formal religious dimension of life has meant an increased trust in science as opposed to religion or tradition.

In food, this has meant that those who have received education that dealt with the physical sciences (probably college) are taking nutritional articles seriously despite the hegemonic domination of print space by advertisements for higher-profit, less healthy items.

The second major development in food change has been the continuing influence of the remains of the counterculture. Whereas the 1960s and 1970s saw a number of food "fads" (organics, Zen macrobiotics, fruititarians), less radical survivors have continued alternative eating traditions. As evidence, one should notice the large number of health food stores that occupy most American cities, as well as those commonly found in suburban malls.

In addition, there currently exists in the United States a well established and health-influenced food cooperative movement (see chapter 1). Ranging from simple buying clubs to full grocery stores, this form of food marketing can subvert the profit principle and lead toward a style of democratic food choice and management. Though not all food co-ops are health food stores, many carry healthy products in response to the kind of people who shop at them. Although they cannot always compete successfully on prices (the economies of scale favor the oligopolies), their alternative products and friendly, democratic atmosphere appeal to many who resist the impersonality and sterility of the typical supermarket.

Further evidence of changes in food consumption that have come to characterize America are the numerous salad bars that have appeared in restaurants. This occurrence crosses the spectrum of fast food, family dining, and even medium-sized rural restaurants across the country. To find such choice ten years ago would have been impossible.

On the other hand, the response of the mainstream "market" has often been one of deception and sleaze. Belasco lists five tactics used by food advertisers:

1. Indeterminant modifiers: words such as "natural" modifying only one ingredient in a package but appearing to modify the entire list

2. Innocence by association: using the words "natural" and "nature" in commercials describing a product that actually contains a number of questionable chemicals

3. Ingredients you'd prefer to avoid: advertising all-natural ingredients but avoiding the percentage (usually very high) of sugar and/or salt content

4. The negative pitch: saying that a product did not have bad contents (like cholesterol) when it does not anyway (e.g., jam)

5. Lying: using "natural" when, in fact, there are questionable chemicals (Belasco 1989, 186)

One expects industry, given its profit and ethical orientations, to attempt to circumvent truth a good deal of the time. For this reason activist groups like the Center for Science in the Public Interest are continually involved in campaigns to bring about truth in advertising.

A Niche Theory of Healthy Eating

When the truth of science and the educated part of the populace come into conflict with the multinational oligopolies that exert hegemonic control over America's food system, what result can we expect? It would be wonderful to think that unbiased scientific truth would prevail. But from the materialist perspective we have championed in this book, this does not appear likely. Instead, we propose a "niche" theory as the most likely option for healthy eating.

While the great masses of Americans exist on fast food and convenience food, there are enough people who care that their food be healthy and better tasting. At one extreme are the epicureans who frequent elite restaurants in big cities and who dine on the best available food. Another group includes the increasing number of vegetarians and consumers of organic food. These range all the way from vegans (eating no flesh or dairy products) through lacto and ovo vegetarians (milk and eggs) to semi-vegetarians who have merely given up red meat.

In addition, a large constituency (it remains for statisticians to measure it) has made many of the healthy changes that are showing up in American supermarkets and restaurants. From the influence of articles on nutrition to the changes that medical doctors have suggested (and prescribed), many people have made some significant changes in their diet.

Taken together, these groups range from the pure vegetarians to the penumbra of people who have made significant changes in their diets. Though they are not the majority of Americans, they certainly are a market that the supermarket chains are beginning to take seriously.

It is conceivable that the changes begun in the American diet will increase in the direction of healthy eating. Generally the working classes imitate the more affluent ones. And as higher education increasingly begins to assimilate (instead of just talk about) the results of scientific nutrition, one might reasonably expect future generations of more nutrition-conscious consumers.

Should this happen, Gene Logsdon has detailed the beginning of the movement of small-scale specialty farms that already are serving demands for high-quality produce and that provide one possible solution to the future of family farms. At present these specialty farms, or "guerilla marketers," are providing such items as "goat cheese, farm raised oysters, hand-pressed cider, hydroponic spinach, stoneground flours from locally grown organic grains, baby lambs . . . pheasant and other once wild game . . . buffalo, . . . and dairy sheep producing Roquefort cheese (Logsdon 1989, 83). In addition, various small farms around the country raise small herds of sheep for homespun yarn, crayfish, catfish, ostriches, llamas, edible squash flowers and fruits, high-quality yogurt, and organic beef. Further, as shown by the Alar scare, Americans are increasingly concerned about harmful chemicals in their foods. And, as we documented in chapter 8, there is considerable research on large-scale organic and low chemical farming.

Government, though its dominant support is for the oligopolies and petrochemical-style farming interests, has shown some interest in alternative approaches. The United States Department of Agriculture (USDA) now has an office for small-scale agriculture. One panel from the conservative Council for Agricultural Science and Technology included organic farm advocates and recommended reduction of agrichemical use (Belasco 1989, 246).

An optimistic scenario for the future would include some possible alliances between groups that would seem to have similar interests. Along with traditional organic growers stand certain croplike wine grapes that can now be grown economically using organic methods, one might explore alliances between organic growers and other concerned groups.

Certainly worth considering is the general health and exercise movement in America. Begun mostly in noncommercial running (though wooed into fancy equipment by the sporting goods manufacturers), this constituency has a natural interest in eating to support bodily health. And the recent scientific/nutrition literature would tend to make them a fertile ground for education and inclusion in nutritional approaches to eating.

Another potential ally is the environmental movement. Joan Gussow quotes a speaker at the Other Economic Summit in Toronto in 1988 saying that "someone is going to produce the materials out of which each of us is made. We urgently need to find a way to make eaters ask themselves whether they want to trust that responsibility to Philip Morris. My message is that we need to enlist environmentalists much more deeply in our cause" (n.d., 21).

A potentially most important ally is the large number of senior citizens who are changing their eating habits. In a recent conversation with a supermarket executive, I mentioned that they had no health food section. He replied that they did have a section for senior citizens, which contained mostly healthy food. Supermarkets are most aware of their potential markets. They also know that, at long last, medical doctors are taking the nutritional literature seriously. Although the extent of change remains to be investigated, senior citizens are cutting down on red meat (McIntosh 1990), increasing fiber (instead of taking laxatives), and getting vitamins from vegetarian sources. The demographics of senior citizens are compelling. As dietary change comes to a generally accepted manner of treatment for seniors, the market for healthy food will expand geometrically.

At the moment, the healthy food and organic farming movements occupy a significant niche in the American food environment. Should they form alliances with the health, environmental, and senior citizens movements, they could become a significantly more influential force in American life.

On the other hand, just as many supermarkets have started organic produce sections, one might expect a counterattack from the oligopolies. Were this to happen, the dynamics of smallness and informality that now characterize the counter-culture of food might be lost in return for some greater degree of healthy food in the mainstream of American life.

Summary

This chapter has focused on successful examples about food. The first part described the successes China, Cuba, and Kerala have provided for developing countries. Through the use of appropriate technology and labor-intensive agriculture, these countries have overcome hunger. And through their commitment to modern socialism, they have ensured a greater equality in food consumption than the capitalist parts of the developing world.

In America, we have still not solved hunger. One-tenth of us use food stamps. Homelessness and hunger abound. The mainstream is characterized by consumption of beef and the fast-food industry. Nevertheless, a substantial body of healthy eaters is exerting pressure on a food system controlled by profit-seeking oligopolistic corporate giants.

Should this constituency of healthy eaters ally with other potentially healthy eaters, we might witness a rebirth of democracy in the food area. Food could be the beginning of decentralization and economic democracy. As sociologist Georg Simmel once wrote: "Eating and drinking, the oldest and intellectually most negligible functions, can form a tie, often the only one; among very heterogenous groups" (1950, 33).

References

Anand, Sudhir, and S.M. Ravikanbur. 1991. "Public Policy and Basic Needs Provision: Intervention and Achievement in Sri Lanka." In *The Political Economy of Hunger*, vol. 3. *Endemic Hunger*, ed. Jean Dreze and Amartya Sen, 52–92. Oxford: Clarendon Press.

Anderson, E. N. 1988. *The Food of China.* New Haven: Yale University Press.

Ash, Roberta. 1972. *Social Movements in America.* Chicago: Markham.

Belasco, Warren J. 1989. *Appetite for Change: How the Counterculture Took on the Food Industry: 1966–1988.* New York: Pantheon.

Benjamin, Media. 1993. Speech at Grand Rapids, Mich., March 17.

Benjamin, Media, Joseph Collins, and Michael Scott. 1984. *No Free Lunch: Food and Revolution in Cuba Today.* San Francisco: Institute for Food and Development Policy.

Berry, Wendall. 1986. "In Defense of the Family Farm." Speech at Kalamazoo College, Kalamazoo, Mich., October 8.

Bettleheim, Bruno. 1950. *Love is Not Enough.* New York: Free Press.

Bread for the World Institute on Hunger and Development. 1990. *Hunger, 1990.* Washington, D.C.: The Institute.

Brewster, Letitia, and Michael Jacobson. 1983. *The Changing American Diet*. Washington, D.C.: Center for Science in the Public Interest.

Brody, Jane. 1987. *Jane Brody's Nutrition Book; Revised and Updated.* New York: Bantam Books.

Clark, Jeff. 1992. "Coop Capitalism." *Down East*, May, 7, 8, 10.

Collins, Joseph. 1982. *What Difference Could a Revolution Make? Food and Farming in the New Nicaragua*. San Francisco: Institute for Food and Development Policy.

Dreze, Jean, and Amartya Sen. 1989. *Hunger and Public Action*. Oxford: Clarendon Press.

———. 1991. "Introduction." In *The Political Economy of Hunger*, vol. 3. *Endemic Hunger*, ed. Jean Dreze and Amartya Sen, 1–15. Oxford: Clarendon Press.

Edgar, Ian. 1987. "The Symbolic Use of Food in a Therapeutic Community." Paper presented at the first annual meeting of the Association for the Study of Food and Society, Grand Rapids, Mich., March.

Franke, Richard W., and Barbara H. Chasin. 1989. *Kerala: Radical Reform as Development in an Indian State: Food First Development, Report No. 6*. San Francisco: Institute for Food and Development Policy.

———. 1990. "Development Without Growth: The Kerala Experiment." *Technology Review*, April, 43–51.

Gussow, Joan Dye. n.d. "Who's Growing You? The Future of Food Production Is on Your Table Today." *Journal of Sustainable Agriculture*, 5–6, 29–31.

Hanes, Phyllis. 1991. "Americans Are Eating Their Veggies—And Liking Them." *Grand Rapids Press*, April 8, D1, D2.

Harwood, Mary Jo. 1987. "Feeding the People." In *Nigaragua Today: Social Services Eight Years after the Revolution*, ed. Don Cooney, 31–37. Kalamazoo: Western Michigan University.

Hinrichs, Cynthia Clare. 1988. "Attaining Food Self-Sufficiency: The Contrasting Cases of India and China." Paper presented at the meeting of the Association for the Study of Food and Society. Chevy Chase, Md., May.

Kandel, R.F., and G.H. Pelto. 1980. "The Health Food Movement: Social Revitalization or Alternative Health Maintenance System." In *Nutritional Anthropology*, eds. Norge Jerome, R.F. Kandel, and G.H. Pelto, 327–364. Pleasantville, N.Y.: Redgrave.

Leung, Joyce. 1983. "China's Multiple Approach to Its Food Problem." In *Controversial Nutrition Policy Issues*, ed. Georgio R. Salimano and Sally A. Lederman, 390–413. Springfield, Ill.: Charles C. Thomas.

Liebman, Bonnie. 1990a. "The Changing American Diet." *Nutrition Action*, May, 8–9.

———. 1990b. "Lessons from China." *Nutrition Action Health Letter*, December, 1, 5–7.

Logsdon, Gene. 1989. "Who Says the Family Farm Is Dead? Welcome to Future Farming's Best Bet." *Utne Reader*, Spring, 82–88.

McIntosh, William Alex. 1990. "Nutritional Risk and Social Relationships." Paper presented at the meeting of the Association for the Study of Food and Society, Philadelphia, June.

Peper Harow. n.d. *Prospectus*. Godalming, England: n.p.

Riskin, Carl. 1991. "Feeding China: The Experience Since 1949." In *The Political Economy of Hunger*, vol. 3, *Endemic Hunger*, ed. Jean Dreze and Amartya Sen, 15–58. London: Clarendon Press.

Rose, M. 1977. "Residential Treatment: A Total Therapy." David Wills Lecture, Mary Sumner Hall, London.

Shenker, Israel. 1992. "Zealous Defender of the Faith Rules a Vegetable Realm." *Smithsonian Magazine*, February, 70–75.

Simmel, Georg. 1950. "Fundamental Problems of Sociology." In *The Sociology of Georg Simmel*, ed. Kurt Wolff, trs. Glencoe, Ill.: The Free Press of Glencoe.

Smil, Vaclav. 1993. *China's Environmental Crisis*. New York: M.E. Sharpe.

Vandermeer, John. 1986. "Moving Towards Independent Agriculture: Nicaragua Struggles in the World Economy." *Science for the People*, January/February, 16–21.

Warnoch, John W. 1987. *The Politics of Hunger*. New York: Methuen.

Whit, William C. 1990. "The Meaning of the Health Food Movement." Paper presented at the World Congress of Sociology, Madrid, July.

Notes

[1] 1.See chapter 9 for a fuller discussion of China's approach to agriculture.

[2] 2.Thanks are due to Medea Benjamin, Joseph Collins and Michael Scott for their *No Free Lunch: Food and Revolution in Cuba Today* (1984). They provide the only really comprehensive and reliable estimation of the Cuban food system.

[3] 3.Some critics have downplayed the significance of Cuba's success in development by citing the Soviet subsidy. The amount cited was about $3 billion annually. In response to this criticism, Warnock wrote, "Puerto Rico receives about $4 billion in aid from the U.S. government each year, four times more per capita than Cuba receives from the USSR" (1987, 228). At present, the USSR subsidy to Cuba has ended.

[4] 4.A discussion of Nicaraguan and North Korean socialist food practices occurs in Warnock 1987. Joseph Collins discusses Nicaragua in *What Difference Could a Revolution Make*? (1982). See also Anand and Kanbar 1991, 59–93.

Appendix: Peper Harow

In the fall of 1988, while on a sabbatical leave in England, I visited Peper Harow, a community of interest to analysts of food and society. I was seeking to illustrate insights contained in a paper by Ian Edgar ("The Symbolic Use of Food in a Therapeutic Community") presented at the first annual meeting of the Association for the Study of Food and Society in 1987.

Peper Harow is a therapeutic community that, as far as I know, is unique in the world. Acting on the insights of food analysts, Peper Harow self-consciously employs food as a therapeutic medium in the process of treating disturbed English adolescents. Like most therapeutic communities, it seeks to take people with feelings of inferiority, anger, hopelessness, and alienation and transform them into healthy persons. And it assumes that by playing by the rules of the community, people will improve.

Unique to Peper Harow are the theoretical insights from the study of food. Specifically, the giving of food is generally an expression of love. And the quality of the meal is indicative of the quality of the love. At Peper Harow much effort has been expended on conveying, through food usage, love. In this sense, food is one way to replace the parental love that many of these older children lacked in their family backgrounds.

Specifically, the areas of food innovation include decor, involvement with regular meals, snacking and feasts. The elevated serving area of the dining room adjoins the kitchen (without doors). Rose writes, "the cutlery and crockery should not be institutionalized, but of a very high standard, offering very satisfying feelings when used. We would have glasses at the meal table at main meals, not imitation plastic glass. We would have fish knives when we had fish, bread knives, cups and saucers."

Furthermore, the furniture was to be of the highest quality: "our table tops were finely made by Robert Thompson . . . the only firm in the country that produces seasoned, solid English oak, two inches thick, which includes adzed surface. Running one's fingers across the rippling surface as one sits at the table is a pleasurable experience indeed!" In addition, the floor and drapes were of heavy material and the room was sound-proofed. The overall effect was to convey the same meanings of acceptance and love that elite dining experiences convey. This even included summer picnics with white tablecloths taking place in English gardens.

Both staff and residents were involved with meal preparation and clean-up. The kitchen area (not a separate room) adjoined the eating space. The kitchen pots were of a luxury quality and the kitchen itself conveyed a message of care.

Novel also for an institutional setting was the snacking procedure. Unlike most institutional snack areas, which are open only at specific times, the "buttery" at Peper Harow was open continually. The message seems clear that if food expresses love, it is unconditional here. As Edgar writes, "Bettleheim found in his Orthogenic school in Chicago that round the clock availability of basic food snacks relieved residents of one of their most deep rooted fears, that of going hungry."

Feasts are also important in conveying love. The *potlatch* in many tribal cultures both demonstrated the affluence of the giver and his concern for those whom she or he invited. Similarly, one generally invites only one's close friends to feast meals. At Peper Harow great attention is paid to their three yearly feasts. Edgar quotes Melvyn Rose, the director, describing a menu for a Christmas feast:

> Open melon with parma ham, followed by a veloute d'Asperge, made with asparagus from our gar-

> dens. The fish course was smoked salmon and was separated from the roast duck and orange sauce with its chestnut barquettes and potatoes by champagne sorbet. In turn, the entree was followed by Christmas pudding, brought in decorated with holly and flaming in brandy, of course together with real rum butter and accompanied by mince pies and cream. The cheeseboard was brought around after that while coffee and cream was served. There were beautiful receptacles on the table.

The entire eating experience at Peper Harow was self-consciously engineered to convey, at conscious and subconscious (cultural) levels, the concern and love of the staff and their small society. Though data on "success" of all psychotherapies are uneven, the staff reported that they felt the intended results were very successful.

Reference

Rose, Melvyn. 1977. "Residential Treatment—A Total Therapy." Lecture. London.

INDEX